Lecture Notes in Mathematics

Donggao Deng · Yongsheng Han

Harmonic Analysis on Spaces of Homogeneous Type

With a Preface by Yves Meyer

Donggao Deng (deceased)
Yongsheng Han
Department of Mathematics and Statistics
Auburn University
Auburn, AL 36849-5310
USA
hanyong@auburn.edu

ISBN 978-3-540-88744-7 ISBN 978-3-540-88745-4 (eBook)
DOI 10.1007/978-3-540-88745-4

Lecture Notes in Mathematics ISSN print edition: 0075-8434
ISSN electronic edition: 1617-9692

Library of Congress Control Number: 2008938190

Mathematics Subject Classification (2000): 35J40, 41A30, 42B20, 42B30, 42C15, 42C40, 43A15, 46F12

Cover design: SPi Publishing Services

Printed on acid-free paper

9 8 7 6 5 4 3 2 1

springer.com

Preface

This book could have been entitled "Analysis and Geometry." The authors are addressing the following issue: Is it possible to perform some harmonic analysis on a set? Harmonic analysis on groups has a long tradition. Here we are given a metric set X with a (positive) Borel measure μ and we would like to construct some algorithms which in the classical setting rely on the Fourier transformation. Needless to say, the Fourier transformation does not exist on an arbitrary metric set.

This endeavor is not a revolution. It is a continuation of a line of research which was initiated, a century ago, with two fundamental papers that I would like to discuss briefly.

The first paper is the doctoral dissertation of Alfred Haar, which was submitted at to University of Göttingen in July 1907. At that time it was known that the Fourier series expansion of a continuous function may diverge at a given point. Haar wanted to know if this phenomenon happens for every orthonormal basis of $L^2[0,1]$. He answered this question by constructing an orthonormal basis (today known as the Haar basis) with the property that the expansion (in this basis) of any continuous function uniformly converges to that function.

Today we know that Haar was the grandfather of wavelets and we also know that wavelet bases offer a powerful and flexible alternative to Fourier analysis. Indeed wavelet bases are unconditional bases of most of the functional spaces we are using in analysis. In other words wavelet expansions offer an improved numerical stability, as compared with Fourier series expansions. One of the goals of this book is to construct wavelets on any metric set equipped with a positive measure which is compatible with the given metric. In this setting we do not have Fourier analysis at our disposal.

The second paper which preluded the authors' endeavor was written in French by Marcel Riesz in 1926. It is entitled "Sur les fonctions conjuguées." The author proves that the Hilbert transform is bounded on $L^p(\mathbb{R})$ when $1 < p < \infty$. The Hilbert transform H is the convolution with $\frac{1}{\pi} p.v. \frac{1}{x}$, which is a distribution. In other words $H(f)(x) = \frac{1}{\pi} p.v. \int \frac{f(y)}{x-y} dy$. The Fourier transform of $H(f)$ is $-i\,\mathrm{sign}(\xi)\widehat{f}(\xi)$ when $\widehat{f}(\xi)$ is the Fourier transform of f. Therefore, H is isometric on $L^2(\mathbb{R})$.

The proof given by Riesz relies on the properties of holomorphic functions F in the unit disc $\mathbb{D}$ of the complex plane. The boundary Γ of $\mathbb{D}$ is the unit circle identified to $[0, 2\pi]$ and functions on Γ can be written as Fourier series. If a holomorphic function F in $\mathbb{D}$ extends to the boundary Γ, then the Fourier series of F on Γ coincides with its Taylor series. Moreover if u is the real part of a holomorphic function F and v is the imaginary part, then v is the Hilbert transform of u on Γ.

To prove his claim, Riesz used the Cauchy formula and the fact that F^p (F raised to the power p) is still holomorphic when p is an integer or when F has no zero in $\mathbb{D}$. This attack was named "complex methods" by Antoni Zygmund.

In the 1950s Alberto Calderón and Zygmund discovered a new strategy for proving L^p estimates. They could not use complex methods anymore since they were interested in operators acting on $L^2(\mathbb{R}^n)$. The operators constructed by Calderón and Zygmund are the famous pseudo-differential operators and soon became one of the most powerful tools in partial differential equations.

Let us sketch the proof of L^p estimates discovered by Calderón and Zygmund. It begins with a lemma which is known as the "Calderón–Zygmund decomposition." It says the following. Let f be any function in $L^1(\mathbb{R}^n)$ and let $\lambda > 0$ be a given threshold. Then f can be split into a sum $u + v$ where $|u|$ is bounded by λ and belongs to $L^2(\mathbb{R}^n)$, while v is oscillating and supported by a set of measure not exceeding $\frac{C}{\lambda}$. As noticed by Joseph Doob, the proof of this lemma is indeed a stopping time argument applied to a dyadic martingale. On the other hand, the Haar basis yields a martingale expansion. Calderón and Zygmund argued as follows. They assumed that the distributional kernel $K(x, y)$ of an operator T satisfies the following conditions: There exists a constant C such that for every $x \in \mathbb{R}^n$ and every $x' \neq x$ one has

$$\int_{|y-x|\geq 2|x'-x|} |K(x', y) - K(x, y)|dy \leq C$$

and there exists a constant C' such that for every $y \in \mathbb{R}^n$ and every $y' \neq y$ one has

$$\int_{|x-y'|\geq 2|y-y'|} |K(x, y') - K(x, y)|dx \leq C'. \qquad (\dagger)$$

Calderón and Zygmund proved a remarkable result. If T is bounded on $L^2(\mathbb{R}^n)$ and if the distributional kernel $K(x, y)$ of T satisfies (†), then for every f in $L^1(\mathbb{R}^n), T(f)$ belongs to weak L^1. There exists a constant C such that for every positive λ the measure of the set of points x for which $|T(f)(x)| > \lambda$ does not exceed $C\frac{\|f\|_1}{\lambda}$. This is optimal, since $f = \delta_{x_0}$ (Dirac mass at x_0) yields

$T(f)(x) = K(x, x_0)$ which belongs to weak L^1 and not to L^1. This theorem follows from the Calderón–Zygmund decomposition. Then the Marcinkiewicz interpolation theorem implies the required L^p estimates for $1 < p \leq 2$. Applying the same argument to the adjoint operator T^*, we obtain the L^p estimates for $2 \leq p < \infty$.

The arguments which were used in these two steps do not rely on Fourier methods; therefore, this scheme easily extends to geometrical settings where the Fourier transformation does not exist. Such generalizations were achieved by Ronald Coifman and Guido Weiss. They discovered that the "spaces of homogeneous type" are the metric spaces to which the Calderón–Zygmund theory extends naturally. A space of homogeneous type is a metric space X endowed with a positive measure μ which is compatible with the given metric in a sense which will be detailed in this book. Roughly speaking, the measure $\mu(B(x,r))$ of a ball centered at x with radius r scales as a power of r.

Coifman and Weiss observed that any bounded operator $T : L^2(X, d\mu) \to L^2(X, d\mu)$ whose distributional kernel satisfies (†)—with $|x - y'| \geq 2|y - y'|$ replaced by $d(x, y') \geq 2d(y, y')$—maps L^1 into weak L^1. That implies L^p estimates for $1 < p \leq 2$. This can be found in the remarkable book *Analyse Harmonique Non- commutative sur Certains Espaces Homogènes* which was published in 1971.

But this does not tell us how to prove the fundamental L^2 estimate. We will return to this issue after a detour.

In the 1960s Calderón launched an ambitious program. He wanted to free the pseudo-differential calculus from the unnecessary smoothness assumptions which were usually required to obtain commutator estimates. The first issue he addressed was the following problem. Let A be the pointwise multiplication by a function $A(x)$ and let T be any pseudo-differential operator of order 1. Can we find a necessary and sufficient condition on A implying that all commutators $[A, T]$ are bounded on $L^2(\mathbb{R}^n)$? This is required for every pseudo-differential operator of order 1 and the particular choices $T_j = \frac{\partial}{\partial x_j}, 1 \leq j \leq n$, show that A must be a Lipschitz function. The other way around is much more difficult and was proved by Calderón in 1965. The proof relies on new estimates on the Hardy space $\mathcal{H}^1(\mathbb{R})$. Calderón proved that the $\mathcal{H}^1$ norm of a holomorphic function F is controlled by the L^1 norm of the Lusin area function of F. This connection between an L^2 estimate and the Hardy space $\mathcal{H}^1$ is the most surprising. An explanation will be given by the $T(1)$ theorem of David and Journé.

This spectacular achievement gave a second life to the theory of Hardy spaces and Charles Fefferman, in collaboration with Elias Stein, proved that the dual of $H^1(\mathbb{R}^n)$ is $BMO(\mathbb{R}^n)$. Here $H^1(\mathbb{R}^n)$ is the real variable version of the Hardy space $H^1(\mathbb{R})$. In other words, H^1 is the subspace of L^1 which is defined by $n+1$ conditions $f \in L^1$ and $R_j f \in L^1$, where $R_j, 1 \leq j \leq n$, are the Riesz transforms.

Calderón conjectured that the Cauchy kernel on a Lipschitz curve Γ is bounded on $L^2(\mathbb{R})$. A Lipschitz curve Γ is the graph of a (real-valued)

Lipschitz function A. The curve Γ admits a parameterization given by $z(x) = x + iA(x), -\infty < x < \infty$, and the Cauchy operator can be written as

$$C(f)(x) = p.v. \frac{1}{\pi i} \int_{-\infty}^{\infty} (z(x) - z(y))^{-1} f(y) dy.$$

If $\|A'\|_\infty < 1$, the Cauchy operator is given by a Taylor expansion $\sum_0^\infty C_n(f)$, where C_n are the iterated commutators between A (the pointwise multiplication with $A(x)$) and $D^n H$. Here, as above, H is the Hilbert transform and $D = -i\frac{d}{dx}$.

In 1977 Calderón used a refinement of the method which was successful for the first commutator and could prove the boundedness of the Cauchy kernel under the frustrating condition $\|A'\|_\infty < \beta$, where β is a small positive number. Guy David combined this result with new real variable methods and got rid of the limitation in Calderón's theorem.

But the main breakthrough came when David and Jean-Lin Journé attacked a much more general problem. They moved to $\mathbb{R}^n$ and studied singular integral operators which are defined by

$$T(f)(x) = p.v. \int K(x, y) f(y) dy,$$

where $K(x, y) = -K(y, x)$, $|K(x, y)| \le C|x - y|^{-n}$, and $|\nabla_x K(x, y)| \le C'|x - y|^{-n-1}$.

They discovered that T is bounded on $L^2(\mathbb{R}^n)$ if and only if $T(1) \in BMO(\mathbb{R}^n)$. Here $T(1)(x) = p.v. \int K(x, y) dy$ and in many situations this calculation is trivial. For instance, when $K_n(x, y) = \frac{(A(x)-A(y))^n}{(x-y)^{n+1}}$ is the n-th commutator,

$$p.v. \int K_n(x, y) dy = -\frac{1}{n} p.v. \int K_{n-1}(x, y) A'(y) dy,$$

which immediately yields Calderón's theorem. Complex methods are beaten by real variable methods and the surprising connection between Hardy spaces and L^2 estimates is explained. Indeed BMO is the dual of H^1.

A spectacular discovery by David, Journé, and S. Semmes is the generalization of the $T(1)$ theorem to spaces of homogeneous type.

This version of the $T(1)$ theorem will receive a careful exposition in this book. It paves the road to a broader program which is the extension to spaces of homogeneous type of the Littlewood–Paley theory. The Littlewood–Paley theory began with the fundamental achievements of J. E. Littlewood and R. E. A. C. Paley.

Let me say a few words on this discovery. We consider the Fourier series $\sum_{-\infty}^{\infty} c_k exp(ikx)$ of a 2π-periodic function $f(x)$ and we define the dyadic blocks $D_j(f)(x), j \in \mathbb{N}$, by

$$D_j f(x) = \sum_{2^j \le |k| < 2^{j+1}} c_k \, exp(ikx).$$

Then the square function $S(f)$ of Littlewood and Paley is defined by

$$S(f)(x) = \Big(\sum_0^\infty |D_j(f)(x)|^2\Big)^{\frac{1}{2}}.$$

Littlewood and Paley proved that we have

$$c_p \|f\|_p \le |c_0| + \|S(f)\|_p \le C_p \|f\|_p$$

when $1 < p < \infty$.

The definition of the square function $S(f)$ was generalized by Elias Stein. Then $L^p[0, 2\pi]$ can be replaced by $L^p(\mathbb{R}^n)$. Jean-Michel Bony used Stein's version of the Littlewood–Paley theory to construct his famous paraproducts. Such paraproducts play a pivotal role in the proof of the $T(1)$ theorem.

The authors of this book show us how to extend the Littlewood–Paley theory to spaces of homogeneous type. This is a key achievement since most of the usual functional spaces admit simple characterizations using the Littlewood–Paley theory.

The last but not the least contribution of the authors is the construction of wavelet bases on spaces of homogeneous type. Once again, wavelets offer an alternative to Fourier analysis. As we know, wavelet analysis can be traced back to a fundamental identity discovered by Calderón. If ψ is a radial function in the Schwartz class with a vanishing integral and if, for $t > 0, \psi_t(x) = t^{-n}\psi(\frac{x}{t})$, then for $f \in L^2(\mathbb{R}^n)$ we have

$$f = c \int_0^\infty f * \widetilde{\psi}_t * \psi_t \frac{dt}{t},$$

where $c > 0$ is a normalizing factor and $\widetilde{\psi}(x) = \overline{\psi}(-x)$. In other words, one computes the wavelet coefficients by

$$W(y, t) = \int f(x) \overline{\psi}_t(x - y) dx$$

and one recovers f through

$$f(x) = c \int_0^\infty \int_{\mathbb{R}^n} W(y, t) \psi_t(x - y) dy \frac{dt}{t}.$$

Everything works as if the wavelets $\psi_{t,y}(x) = t^{-n/2}\psi(\frac{x-y}{t})$ were an orthonormal basis of $L^2(\mathbb{R}^n)$. Indeed, orthonormal wavelet bases exist. There exist $2^n - 1$ functions $\psi_\epsilon \in \mathscr{S}(\mathbb{R}^n), \epsilon \in F, \#F = 2^n - 1$, such that the functions $\psi_\epsilon(x) = 2^{\frac{nj}{2}} \psi_\epsilon(2^j x - k), j \in \mathbb{Z}, k \in \mathbb{Z}^n, \epsilon \in F$, are an orthonormal basis of $L^2(\mathbb{R}^n)$.

The authors succeeded in generalizing the construction of wavelet bases to spaces of homogeneous type; however, wavelet bases are replaced by frames, which in many applications offer the same service.

One is amazed by the dramatic changes that occurred in analysis during the twentieth century. In the 1930s complex methods and Fourier series played a seminal role. After many improvements, mostly achieved by the Calderón–Zygmund school, the action takes place today on spaces of homogeneous type. No group structure is available, the Fourier transform is missing, but a version of harmonic analysis is still present. Indeed the geometry is conducting the analysis.

Donggao Deng passed away after completing a preliminary version of this book. In his last moments he knew his efforts were not in vain and that his collaboration with Yongsheng Han would eventually lead to this remarkable treatise.

China 2007 *Yves Meyer*

Contents

Introduction

As Y. Meyer wrote in [M2]: "At the beginning of the 1980's, many scientists were already using 'wavelets' as an alternative to traditional Fourier analysis. This alternative gave grounds for hoping for simpler numerical analysis and more robust synthesis of certain transitory phenomena." He also wrote: "To mention only the most striking, R. Coifman and G. Weiss invented the 'atoms' and 'molecules' which were to form the basic building blocks of various function spaces, the rules of assembly being clearly defined and easy to use. Certain of these atomic decompositions could, moreover, be obtained by making a discrete version of a well-known identity, due to A. Calderón, in which 'wavelets' were implicitly involved. That identity was later rediscovered by Morlet and his collaborators." Y. Meyer further wrote: "These separate investigations had such a 'family resemblance' that it seemed necessary to gather them together into a coherent theory, mathematically well-founded and, at the same time, universally applicable."

Today we know that this coherent theory is wavelet analysis. This theory played and will, doubtless, play an important role in many different branches of science and technology. Wavelet analysis provides a simpler and more efficient way to analyze those functions and distributions that have been studied by use of Fourier series and integrals. But, however, Fourier analysis still plays a key role in constructing the orthonormal bases of wavelets.

Fourier analysis also plays an essential role in the study of the boundedness of convolution operators. In the 1950's when Calderón and Zygmund systematically studied convolution operators appearing in elliptic partial differential equations, they developed the theory of Calderón-Zygmund convolution operators. The continuity of these Calderón-Zygmund convolution operators on $L^2(\mathbb{R}^n)$ follows immediately from the Fourier transform. The boundedness of such operators on $L^p(\mathbb{R}^n), 1 < p < \infty$, is then obtained by the so-called real variable method of Calderón and Zygmund. This real variable method, in general, still applies to non-convolution operators whenever kernels of those non-convolution operators satisfy the regularity conditions of Calderón and Zygmund and the continuity of those non-convolution operators on $L^2(\mathbb{R}^n)$

D. Deng and Y. Han, *Harmonic Analysis on Spaces of Homogeneous Type*, Lecture Notes in Mathematics 1966,

has been known. These non-convolution operators are now called Calderón-Zygmund operators. Therefore, the L^2 boundedness becomes the core of the boundedness of Calderón-Zygmund operators on other functional spaces. Obviously, the Fourier transform is no longer available to get the L^2 continuity for non-convolution operators. The $T1$ theorem, that is, the necessary and sufficient conditions of the L^2 boundedness of non-convolution operators, as a substitute of the Fourier transform, was finally proved by G. David and J. L. Journé ([DJ]). In the original proof of the $T1$ theorem on $\mathbb{R}^n$, the Fourier transform was not used explicitly, but implicitly.

Today we have reached a better understanding of the connections between wavelets and Calderón-Zygmund operators. Indeed any Calderón-Zygmund operator T satisfying $T(1) = T^*(1) = 0$ is almost diagonal in any orthonormal wavelet bases. This applies, for example, to the Hilbert transform and solves the problem raised by Antoni Zygmund. Conversely an operator which is diagonal or almost diagonal in a wavelet basis is such a Calderón-Zygmund operator. A question arises: How does one prove the $T1$ theorem on spaces of homogeneous type? Here spaces of homogeneous type were introduced in the 1970's by R. Coifman and G. Weiss in order to develop the theory of Calderón-Zygmund operators in a more general setting. There are, however, no translations or dilations, and no analogue of the Fourier transform or convolution operation on general spaces of homogeneous type. In 1985, using Coifman's idea on decomposition of the identity operator, G. David, J. L. Journé and S. Semmes developed the Littlewood-Paley analysis on spaces of homogeneous type and used it to give a proof of the $T1$ theorem on this general setting. Before explaining Coifman's idea and the Littlewood-Paley analysis on spaces of homogeneous type, let us first describe the Littlewood-Paley analysis on the standard case of $\mathbb{R}^n$.

Two versions of the Littlewood-Paley analysis exist. The first one is the most familiar and can be traced back to the early thirties. It amounts to splitting a Fourier series expansion into dyadic blocks $\Delta_j(f)$. We then have $f = \sum_j \Delta_j(f)$ and most functional spaces are characterized by size estimates on $\Delta_j(f)$. For instance f is Hölder α if and only if $\|\Delta_j(f)\|_\infty \leq C2^{-j\alpha}$.

If one moves from Fourier series expansion to problems concerning functions on $\mathbb{R}^n$, the familiar Littlewood-Paley analysis can be defined by the following way. One starts with a bump function ϕ in the Schwartz class $\mathscr{S}(\mathbb{R}^n)$ and one assumes the following properties (a) $\int \phi(x)dx = 1$ and (b) $\int x^\alpha \phi(x)dx = 0$ for $|\alpha| \geq 1$. It often suffices to have (b) when $1 \leq |\alpha| \leq N$ where N is related to the functional space one would like to describe.

Then we write $\phi_j(x) = 2^{nj}\phi(2^j x), j \in \mathbb{Z}$, and the convolution operator S_j are defined by $S_j(f) = f * \phi_j$.

When we let S_j act on $L^2(\mathbb{R}^n)$, we have

$$S_j \to I \ \ (j \to +\infty), \ \ S_j \to 0 \ \ (j \to -\infty)$$

which implies

$$I = \sum_{-\infty}^{\infty} \Delta_j, \quad \Delta_j = S_{j+1} - S_j.$$

This is the first version of the Littlewood-Paley analysis.

One has $\Delta_j(f) = f * \psi_j, \psi_j(x) = 2^{nj}\psi(2^j x)$ and $\psi(x) = 2^n\phi(2x) - \phi(x)$. One should observe that $\psi \in \mathscr{S}(\mathbb{R}^n)$ and $\int x^\alpha \psi(x)dx = 0$ for all $\alpha \in \mathbb{N}^n$. Therefore ψ is a wavelet.

The second version of the Littlewood-Paley analysis is referred to as Calderón's identity. Instead of requiring that $f = \sum\limits_{j=-\infty}^{\infty} \Delta_j(f)$, one instead requires that $\|f\|_2^2 = \sum\limits_{j=-\infty}^{\infty} \|\Delta_j(f)\|_2^2$ which is equivalent to Calderón's identity ([C1])

$$I = \sum_{-\infty}^{\infty} \Delta_j^* \Delta_j. \tag{0.1}$$

If, as above, $\Delta_j(f) = f * \psi_j$, then (0.1) is equivalent to

$$1 = \sum_{-\infty}^{\infty} |\widehat{\psi}(2^{-j}\xi)|^2, \qquad \xi \neq 0.$$

Unfortunately the construction of ψ heavily relies on the full power of Fourier analysis and cannot be achieved on a space of homogeneous type.

To circumvent this drawback and develop a form of Littlewood-Paley analysis on a space of homogeneous type, G. David, J. L. Journé and S. Semmes used an idea of Coifman and constructed a remarkable decomposition of the identity operator. Let $\{D_k\}$ be a family of operators whose kernels satisfy certain size, smoothness, cancellation conditions, and the nondegeneracy condition, see Chapter 1 below for more details, such that

$$I = \sum_{k \in \mathbb{Z}} D_k \tag{0.2}$$

on L^2.

Coifman's idea consists in rewriting (0.2) as

$$I = \sum_{k \in \mathbb{Z}} D_k \sum_{l \in \mathbb{Z}} D_l = \sum_{|l|>N} \sum_{k \in \mathbb{Z}} D_{k+l} D_k + \sum_{k \in \mathbb{Z}} \Big(\sum_{|l| \leq N} D_{k+l} \Big) D_k = R_N + T_N,$$

where $R_N = \sum\limits_{|l|>N} \sum\limits_{k \in \mathbb{Z}} D_{k+l} D_k, T_N = \sum\limits_{k \in \mathbb{Z}} D_k^N D_k$ with $D_k^N = \sum\limits_{|l| \leq N} D_{k+l}$, and N is a fixed large integer.

In [DJS] it was proved, using Cotlar-Stein-type lemma and Calderón-Zygmund theory, that for N sufficiently large, T_N^{-1} exists on L^2 and is bounded on $L^p, 1 < p < \infty$. This in turn permitted to establish the Littlewood-Paley estimates on $L^p, 1 < p < \infty$. Indeed for $1 < p < \infty$ there exists a constant C_p, such that

$$C_p^{-1}\|f\|_p \leq \left\|\left\{\sum_{k\in\mathbb{Z}} |D_k(f)|^2\right\}^{\frac{1}{2}}\right\|_p \leq C_p\|f\|_p.$$

The key step in [DJS] was to show that R_N is a Calderón-Zygmund operator with norm at most $C2^{-N\delta}$ for some $\delta > 0$. The essential ingredient used in [DJS] is the following identity

$$I = T_N^{-1}T_N = \sum_{k\in\mathbb{Z}} T_N^{-1}D_k^N D_k = T_N T_N^{-1} = \sum_{k\in\mathbb{Z}} D_k D_k^N T_N^{-1}. \tag{0.3}$$

In the standard case of $\mathbb{R}^n$, Calderón's identity (0.1) provides a continuous version of wavelet expansions. However, the identities (0.3) cannot directly provide such a wavelet expansion on a space of homogeneous type. The difference between the identities (0.3) on a space of homogeneous type and Calderón's identity (0.1) on $\mathbb{R}^n$ is the presence of the operator T_N^{-1}. The results given in [DJS] only say that T_N^{-1} is bounded uniformly on $L^2(\mathbb{R}^n)$, or more generally, on $L^p(\mathbb{R}^n), 1 < p < \infty$. Therefore one did not know if the identities (0.3) could provide a continuous version of wavelet expansions on spaces of homogeneous type.

Then one wonders if Calderón's identity can be obtained in the general context of a space of homogeneous type where we do not have a Fourier transformation at our disposal. One of the main goals of this book is to build a wavelet analysis on space of homogeneous type. To be precise, we will prove some Calderón's identities and use them to write wavelet expansions of functions and distributions on a space of homogeneous type. Moreover these wavelet expansions will reflect the local or global regularity of functions and distributions as in the standard case.

To achieve this goal, we will first define test functions and what we call *wavelet spaces* on a space of homogeneous type. Then we will prove a new "$T1$ theorem", namely the boundedness on *wavelet spaces* of a certain class of Calderón-Zygmund operators whose kernels satisfy some additional second order smoothness condition.

One then checks that the operator R_N, as mentioned above, belongs to this class of Calderón-Zygmund operators with an operator norm at most $C2^{-N\delta}$ for some $\delta > 0$. This implies that T_N^{-1} is bounded on these *wavelet spaces* when N is large.

Let $\widetilde{D}_k = T_N^{-1}D_k^N$ and $\widetilde{\widetilde{D}}_k = D_k^N T_N^{-1}$. Then we will prove that the kernels of $\widetilde{D}_k$ and $\widetilde{\widetilde{D}}_k$ are wavelets in the following sense. We say that a function $K(x,y)$ defined on the product space $X \times X$ is a wavelet if it is a wavelet in y for each frozen x and vice versa. The precise definition of wavelets is to be found in Definition 2.2 below. Therefore we can rewrite (0.3) as

$$I = \sum_{k \in \mathbb{Z}} \widetilde{D}_k D_k = \sum_{k \in \mathbb{Z}} D_k \widetilde{\widetilde{D}}_k, \tag{0.4}$$

which provides a new Calderón's identity and hence a continuous version of wavelet expansions on space of homogeneous type.

Finally, we will replace (0.4) by a discrete version of a wavelet expansion, obtained by using the dyadic cubes of M. Christ on spaces of homogeneous type ([Ch2]) and the boundedness of operators on *wavelet spaces*. These wavelets are not an orthonormal basis, but instead are a frame $\psi_\lambda, \lambda \in \Lambda$. It means that the wavelet expansion of a function $f \in L^2(X, d\mu)$ into a wavelet series will be given by

$$f(x) = \sum_{\lambda \in \Lambda} \alpha(\lambda)\psi_\lambda(x)$$

where

$$\|f\|_2 \approx \left\{ \sum_{\lambda \in \Lambda} |\alpha(\lambda)|^2 \right\}^{\frac{1}{2}}$$

and where the wavelet coefficients $\alpha(\lambda)$ are given by

$$\alpha(\lambda) = \int f(x)\widetilde{\psi}_\lambda(x) d\mu(x). \tag{0.5}$$

We will prove the following result. The dual wavelets $\widetilde{\psi}_\lambda$ are sharing with ψ_λ the same localization, smoothness and vanishing integral properties. This will imply that most functional spaces can be characterized by simple size properties of the wavelet coefficients in (0.5). These wavelet expansions and characterizations of functional spaces will constitute the heart of this book.

Up to now, even though one does not know how to construct orthonormal wavelet bases on general spaces of homogeneous type, these wavelet expansions will provide us with a new and universally applicable tool at our command: "everything that comes to hand", which can be used to carry out many important results on $\mathbb{R}^n$ to spaces of homogeneous type.

In the case of a space of homogeneous type, the Hölder regularity of the modified distance function $\rho(x, y)$, see Chapter 1 for details, is replacing the differential structure in the standard situation of $\mathbb{R}^n$ or of a differentiable manifold. The construction of wavelets will follow from the properties of $\rho(x, y)$. We do not have other tools at our disposal to prove the existence of wavelet bases on spaces of homogeneous type.

The contents of this book may be roughly summarized as follows.

In the first chapter, we restate the major notation and basic results of the theory of Calderón-Zygmund operators on spaces of homogeneous type. We describe the Littlewood-Paley analysis and prove the $T1$ theorem of G. David, J. L. Journé and S. Semmes on spaces of homogeneous type.

We then define test functions and *wavelet spaces* on spaces of homogeneous type in the second chapter (see Definition 2.2). We prove one of main results in this book, namely the boundedness of Calderón-Zygmund operators whose kernels satisfy some additional smoothness condition, on *wavelet spaces*. This result will be the main tool used in Chapter 3.

Chapter 3 is addressed to all kinds of wavelet expansions on spaces of homogeneous type. For reader's convenience some few facts about frames on a Hilbert space are included.

Chapter 4 is devoted to the study of functional spaces on spaces of homogeneous type in terms of wavelet coefficients. These spaces include $L^p, 1 < p < \infty$, the generalized Sobolev spaces $\dot{L}^{p,s}$, the Hardy spaces H^p, BMO and the Besov spaces. The $T1$ type theorems on these spaces are presented.

In recent years it has been known that central results of the classical theory of Calderón-Zygmund operators hold true in very general situations where the standard doubling condition on the underlying measure is not needed. Metric spaces where this happens are now called non homogeneous spaces. It came as a great surprise. Indeed the theory of spaces of homogeneous type were so much adapted to the theory of Calderón-Zygmund operators that everyone believed they were essentially the right context. In the last chapter, Chapter 5, we will outline the Littlewood-Paley analysis on non homogeneous space obtained by X. Tolsa. It is still based on Coifman's decomposition of the identity operator, Although we do not know if there exists a wavelet expansion on a non homogeneous space, we will show that Tolsa's Littlewood-Paley analysis on $L^p, 1 < p < \infty$, still works for the study of new functional spaces, namely the Besov spaces on non homogeneous spaces.

This book is based on some recent work by the authors and their collaborators. This research project started twenty years ago: The first author was invited by R. Coifman to visit Yale University and in the meantime, the second author was invited by G. Weiss to visit Washington University. At that time, the authors planned to build a Littlewood-Paley analysis and then use it to develop a wavelet analysis on spaces of homogeneous type. R. Coifman kindly suggested a kind of identity to the first author, as a key tool to achieve the above aim. Meanwhile, the second author was discussing with G. Weiss on the first manuscript on orthonormal wavelet bases written by Y. Meyer.

Partial material of this book is taken from the lecture notes of graduate students classes given by the second author in the Fall semester 1999 at Washington University in St. Louis and the lectures given at Zhongshan University and Beijing Normal University in China. The first author thanks R. Coifman for his invitation and suggestions. The second author thanks G. Weiss for his invitation in 1999 and for his continuous encouragement and support during many years. The second author would also like to thank E. Sawyer and D. Yang for fruitful collaborations that they have had throughout the years. A thanks goes to Ji Li, Chin-Cheng Lin and Kunchuan Wang.

Without their help, this book would not be presented by this final version in LaTex.

This book was supported by the NSF in China. The first author thanks the NSF in China for its support. This book has been written during the second author's visit at Zhongshan University. He would like to thank Zhongshan University for their warm hospitality and support.

Chapter 1
Calderón-Zygmund Operator on Space of Homogeneous Type

1.1 Introduction

In the 1970's, in order to extend the theory of Calderón-Zygmund singular integrals to a more general setting, R. Coifman and G. Weiss introduced certain topological measure spaces which are equipped with a metric which is compatible with the given measure in a sense which will be detailed in this chapter. These spaces are called spaces of homogeneous type. In this chapter we present the major notational conventions and basic results of the theory of Calderón-Zygmund operators on spaces of homogeneous type. As we already noticed, it becomes indispensable to have a criterion for L^2 continuity, without which the theory collapses like a house built on sandy beach. One such criterion is the $T1$ theorem of G. David, J. L. Journé and S. Semmes on spaces of homogeneous type. Before proving the $T1$ theorem of G. David, J. L. Journé and S. Semmes, we will explain the Littlewood-Paley analysis on spaces of homogeneous type, which, based on Coifman's idea on decomposition of the identity operator, was developed by the above authors. The Littlewood -Paley analysis on spaces of homogeneous type becomes a starting point to provide wavelet expansions of functions and distributions. This will be addressed in Chapter 3.

1.2 Definition of Calderón-Zygmund Operators on Spaces of Homogeneous Type

A *quasi-metric* δ on a set X is a function $\delta : X \times X :\longrightarrow [0, \infty]$ satisfying

$$\delta(x, y) = 0 \text{ if and only if } x = y, \tag{1.1}$$

$$\delta(x, y) = \delta(y, x) \text{ for all } x, y \in X, \tag{1.2}$$

D. Deng and Y. Han, *Harmonic Analysis on Spaces of Homogeneous Type*, Lecture Notes in Mathematics 1966,

and there exists a constant $A < \infty$ such that for all $x, y, z \in X$,

$$\delta(x,z) \leq A[\delta(x,y) + \delta(y,z)]. \tag{1.3}$$

Any quasi-metric defines a topology, for which the balls $B(x,r) = \{y \in X : \delta(x,y) < r\}$ form a base. But the balls need not be open when $A > 1$.

It is not difficult to see that the constant A in (1.3) must be ≥ 1.

We now state the definition of a space of homogeneous type. This definition was introduced by R. Coifman and G. Weiss.

Definition 1.1 ([CW1]). A *space of homogeneous type* (X, δ, μ) is a set X together with a quasi-metric δ and a nonnegative measure μ on X such that $0 < \mu(B(x,r)) < \infty$ for all $x \in X$ and all $r > 0$, and so that there exists $A' < \infty$ such that for all $x \in X$ and $r > 0$,

$$\mu(B(x,2r)) \leq A'\mu(B(x,r)). \tag{1.4}$$

Here μ is assumed to be defined on a σ-algebra which contains all Borel sets and all balls $B(x,r)$ for $x \in X$ and $r > 0$. The property of the measure μ in (1.4) is called the doubling condition. We suppose that $\mu(\{x\}) = 0$ for all $x \in X$. Macias and Segovia have shown ([MS] Theorems 2 and 3) that in this case, the functional

$$\rho'(x,y) = \inf\{\mu(B) : B \text{ are balls containing } x \text{ and } y\} \tag{1.5}$$

is a quasi-metric on X yielding the same topology as δ. Moreover, there is a quasi-metric ρ equivalent to ρ' in the sense that $C^{-1}\rho'(x,y) \leq \rho(x,y) \leq C\rho'(x,y)$ for some constant $C > 0$ and for all $x, y \in X$. Moreover there are a constant $C > 0$, and $0 < \theta < 1$ such that

$$C^{-1}r \leq \mu(B(x,r)) \leq Cr \tag{1.6}$$

for all $x \in X, r > 0$, and

$$|\rho(x,y) - \rho(x',y)| \leq C(\rho(x,x'))^{\theta}[\rho(x,y) + \rho(x',y)]^{1-\theta} \tag{1.7}$$

for all x, x' and $y \in X$.

It is easy to verify, by (1.7), that the balls associated to ρ are open sets. Moreover, there is a positive constant M such that $d(x,y) = (\rho(x,y))^{\frac{1}{M}}$ is equivalent to a metric on $X \times X$.

We would like to emphasize that in (1.6), and for the remainder of this book, all balls are ρ-balls defined by $B(x,r) = \{y \in X : \rho(x,y) < r\}$, where ρ satisfies the estimate (1.7).

From (1.6), we obtain, by decomposing into annuli for example, the often used following estimates

$$\int_{B(x,r)} (\rho(z,x))^{\epsilon-1} d\mu(z) \approx r^{\epsilon} \text{ and } \int_{B(x,r)^c} (\rho(z,x))^{-\epsilon-1} d\mu(z) \approx r^{-\epsilon} \tag{1.8}$$

for all $\epsilon > 0$.

The Hölder spaces on a space of homogeneous type are defined as follows.

Definition 1.2. Let $C_0^\eta(X), \eta > 0$, be the space of all continuous functions on X with compact support such that

$$\|f\|_{C^\eta} = \sup_{x \neq y} \frac{|f(x) - f(y)|}{\rho(x,y)^\eta} < \infty. \tag{1.9}$$

We would like to remark that for $0 < \eta \leq \theta, C_0^\eta(X)$ is not empty (i.e. reduced to $\{0\}$). To see this, we let $f(x)$ be a C^1 function defined on $\mathbb{R}$ with a compact support. Define $g(x) = f(\rho(x, x_0))$ for any fixed $x_0 \in X$. It is easy to check that g is a function in $C_0^\eta(X)$ with $0 < \eta \leq \theta < 1$.

It is well known that the homogeneous Banach space $C^\beta(\mathbb{R})$ of Hölder functions with exponent β has the following properties (a) if $0 < \beta < 1$, $\dot{C}^\beta(\mathbb{R})$ is isomorphic to ℓ^∞, (b) if $\beta = 1$, the Zygmund class is isomorphic to ℓ^∞ and the class of Lipschitz functions is isomorphic to $L^\infty(\mathbb{R})$.

It implies that the dual space of $C^\beta(\mathbb{R})$ is not a functional space. Indeed the dual space of $\ell^\infty(\mathbb{Z})$ is not a space of sequences. The same objection applies to $C_0^\beta(\mathbb{R})$ or $C_0^\beta(X)$. This, however, can be solved. It suffices to replace $C^\beta(\mathbb{R})$ by the closure $\dot{C}^\beta(\mathbb{R})$ for the C^β norm of $C^\gamma(\mathbb{R})$ where $\gamma > \beta$. This closure does not depend on γ.

Following this argument we define the function space $\dot{C}_0^\eta(X)$ as the closure for the $C_0^\eta(X)$ norm of functions in $C_0^\beta(X)$ where $\eta < \beta$, and let $(\dot{C}_0^\eta(X))'$ be dual space of $\dot{C}_0^\eta(X)$. The following theorem implies that these new spaces do not depend on β.

Theorem 1.3. *The two following properties of $f \in C^\alpha(X)$ are equivalent ones*

(1) $f \in \dot{C}_0^\alpha(X)$, the closure of functions in $\dot{C}_0^\beta(X), \beta > \alpha$, with respect to the C^α norm;

(2) $\lim_{\rho(x,y)\to 0, or\, \rho(x,y)\to\infty} (f(x) - f(y))[\rho(x,y)]^{-\alpha} = 0$

and this limit is uniform in $(x, y) \in X \times X$.

The implication (1) $\Rightarrow$ (2) is easy. The results given in [MS] can be used to show the other implication.

Before describing the theory of Calderón-Zygmund operators on spaces of homogeneous type, we give several examples of spaces of homogeneous type ([CW2] and [Ch1]).

(1) $X = \mathbb{R}^n$, $\rho(x,y) = |x - y| = \Big(\sum_{j=1}^n (x_j - y_j)^2\Big)^{\frac{1}{2}}$ and μ equals Lebesgue measure.

(2) $X = \mathbb{R}^n$, $\rho(x,y) = \sum_{j=1}^n (x_j - y_j)^{\alpha_j}$, where $\alpha_1, \alpha_2, ..., \alpha_n$ are positive numbers, not necessarily equal, and μ equals Lebesgue measure (this distance is called nonisotropic).

(3) $X = [0,1)$, $\rho(x,y)$ is the length of the smallest dyadic interval containing x and y, and μ is Lebesgue measure.

(4) X is the boundary of a Lipschitz domain in $\mathbb{R}^n$, ρ is the Euclidean distance and μ is the harmonic measure or the Lebesgue surface measure.

(5) $X = \mathbb{R}^+ = \{r \in \mathbb{R} : r \geq 0\}$, $d\mu(r) = r^{n-1}dr$, ρ is the usual distance.

(6) Any C^∞ compact Riemannian manifold with the Riemannian metric and volume.

(7) Let $n \geq 1$ be an integer and $d \in (0,n]$ a real number. Let $E \subset \mathbb{R}^n$ be a closed subset whose d-dimensional Hausdorff measure $\lambda_d(E)$ is finite and positive. Suppose it happens that there exists $C < \infty$ so that for every $x \in E$ and $r > 0$, $C^{-1}r^d \leq \lambda_d(E \cap B(x,r)) \leq Cr^d$. Then equipped with the Euclidean metric and $\mu = \lambda_d$, E is a space of homogeneous type.

(8) Let G be a nilpotent Lie group with a left-invariant Riemannian metric and μ is the induced measure.

(9) Let Ω be an open set in $\mathbb{R}^n$ and let $X_1, \cdots, X_k$ be C^∞ vector fields in Ω. Suppose that $\{X_j\}$ satisfy the condition of Hörmander. It means that together with all their commutators of all orders, $\{X_j\}$ span the tangent space to $\mathbb{R}^n$ at each $x \in \Omega$. We say that a Lipschitz curve $\gamma : [0,r] \to \Omega$ is admissible if for almost every t, $\frac{d\gamma}{dt} = \sum\limits_{j=1}^{k} c_j(t)X_j(\gamma(t))$ where $\sum |c_j(t)|^2 \leq 1$. Define $\rho(x,y)$ to be the infimum of the set of all r for which there exists an admissible curve γ with $\gamma(0) = x$ and $\gamma(r) = y$. It was proved that such an admissible curve exists for any $x, y \in \Omega$ (provided Ω is connected). Then (Ω, ρ) with the Lebesgue measure becomes a space of homogeneous type. See [NSW] for a detailed discussion of various concrete examples.

(10) In an open subset Ω of $\mathbb{R}^n$ let $A(x) = (a_{i,j}(x))_{i,j=1}^n$ be a C^∞ family of symmetric matrices with real-valued entries, nonnegative in the sense that $\sum\limits_{i,j=1}^{n} a_{i,j}(x)t_it_j \geq 0$ for all x and $t \in \mathbb{R}^n$. A vector $v \in \mathbb{R}^n$ is said to be subunit at x if for every $t \in \mathbb{R}^n$,

$$\left|\sum_j t_j v_j\right|^2 \leq \sum_{i,j=1}^{n} a_{i,j}(x)t_it_j.$$

A Lipschitz curve $\gamma : [0,r] \to \mathbb{R}^n$ is said to be admissible if $d\gamma(s)/ds$ is subunit for almost every s. Define $\rho(x,y)$ to be infinite if there exists no admissible path γ joining x to y, and otherwise $\rho(x,y) = \inf\{r :$ there exists an admissible curve γ satisfying $\gamma(0) = x$ and $\gamma(r) = y\}$. If there exist $\epsilon > 0$ and $C < \infty$ such that $R(x,y) \leq C|x-y|^\epsilon$ for all x and y close to x, then (Ω, ρ) becomes a space of homogeneous type when equipped with Lebesgue measure.

(11) When X is the boundary of a smooth and bounded pseudo-convex domain in $\mathbb{C}^n$ one can introduce a nonisotropic quasi-distance that is related to the complex structure in such a way that we obtain a space of homogeneous type by using Lebesgue surface measure. For example, if X is the surface of

the unit sphere $\sigma_{2n-1} = \left\{z \in \mathbb{C}^n : z \cdot \bar{z} = \sum_{j=1}^{n} z_j \bar{z}_j = 1\right\}$, the nonisotropic distance is given by $d(z,w) = |1 - z \cdot \bar{w}|^{\frac{1}{2}}$.

(12) Let U be a bounded open subset of $\mathbb{C}^2$ with C^∞ boundary. At any $z \in \partial U$ the vector space of linear combinations $a_1 \frac{\partial}{\partial \bar{z}_1} + a_2 \frac{\partial}{\partial \bar{z}_2}$ belonging to the complexified tangent space to ∂U has dimension 1 over $\mathbb{C}$. Fix a nonvanishing C^∞ complex vector field $\bar{L}_{\bar{z}} = a_1 \frac{\partial}{\partial \bar{z}_1} + a_2 \frac{\partial}{\partial \bar{z}_2}$ which is tangent to ∂U at every point z in some open set $V \subset \partial U$. Write $\bar{L} = X + iY$ where X, Y are real vector fields. If U is strictly pseudo-convex then the pair X, Y satisfies the condition of Hörmander and V becomes a space of homogeneous type .

We now introduce the maximal function of Hardy and Littlewood.

Definition 1.4. Suppose $f \in L^1_{loc}(X)$. The Hardy-Littlewood maximal function of f is defined by

$$Mf(x) = \sup_{r>0} \frac{1}{\mu(B(x,r))} \int_{B(x,r)} |f(y)| d\mu(y).$$

Theorem 1.5 ([CW1]). *M is bounded on $L^p(X,\mu)$ for $1 < p \leq \infty$, and is of weak type $(1,1)$.*

Indeed, the Vitali-type covering lemma still holds on spaces of homogeneous type.

Lemma 1.6. *Let $K \subseteq X$ be a compact set. Let $\{B(x_\alpha, r_\alpha)\}$ be a collection of open balls that cover K. Then there is a subcollection $\{B(x_{\alpha_j}, r_{\alpha_j})\}$ that is pairwise disjoint and such that $\{B(x_{\alpha_j}, Cr_{\alpha_j})\}$ still cover K, where $C = 2A^2 + A$.*

Proof. Since K is compact, we may suppose that the collection $\{B(x_\alpha, r_\alpha)\}$ is a finite collection of balls $\{B(x_k, r_k)\}$. Now choose a ball $\{B(x_{k_1}, r_{k_1})\}$ of greatest radius. Then choose $\{B(x_{k_2}, r_{k_2})\}$ from among those balls such that it has greatest possible radius and is disjoint from $\{B(x_{k_1}, r_{k_1})\}$.

Repeating this procedure, the process must eventually stop since the collection is finite. We claim that this new collection satisfies the requirement of Lemma 1.6. By the construction, all balls in this collection are pairwise disjoint. We only need to check that $\{B(x_{k_j}, Cr_{k_j})\}$ cover K for $C = 2A^2 + A$. It suffices to show that $\{B(x_{k_j}, Cr_{k_j})\}$ cover the original collection $\{B(x_k, r_k)\}$. Take one of the original balls $B(x_k, r_k)$ and assume it is not one of the selected balls. Then, by the construction, there is a first selected ball $B(x_{k_p}, r_{k_p})$ that intersects it with $r_{k_p} \geq r_k$. We claim that $B(x_k, r_k) \subseteq B(x_{k_p}, Cr_{k_p})$. To see this, let $z \in B(x_k, r_k) \cap B(x_{k_p}, r_{k_p})$ and let $x \in B(x_k, r_k)$ be arbitrary. Then

$$\rho(x, x_{k_p}) \leq A[\rho(x,z) + \rho(z, x_{k_p})] \leq A[A(\rho(x, x_k) + \rho(x_k, z)) + r_{k_p}]$$

$$\leq A[A(r_k + r_k) + r_{k_p}] \leq [2A^2 + A] r_{k_p}$$

which shows $B(x_k, r_k) \subseteq B(x_{k_p}, Cr_{k_p})$ with $C = 2A^2 + A$. ∎

We also have a Calderón-Zygmund decomposition on spaces of homogeneous type.

Theorem 1.7 ([CW1]). *Let $f \in L^1(X)$ and $\alpha > 0$, and assume that $\mu(X) > \alpha^{-1}\|f\|_1$. Then f can be decomposed as $f = g + b$ where*

$$\|g\|_2^2 \leq C\alpha\|f\|_1,$$

$$b = \sum_j b_j,$$

where each b_j is supported on some ball $B(x_j, r_j)$,

$$\int b_j(x)d\mu(x) = 0,$$

$$\|b_j\|_1 \leq C\alpha\mu(B(x_j, r_j)),$$

$$\sum_j \mu(B(x_j, r_j)) \leq C\alpha^{-1}\|f\|_1.$$

We now introduce Calderón-Zygmund operator on spaces of homogeneous type.

Definition 1.8 ([CW1]). A continuous function $K : X \times X\backslash\{x = y\} \to \mathbb{C}$ (the complement of the diagonal in $X \times X$) is said to be a Calderón-Zygmund singular integral kernel if there exist $0 < \epsilon \leq \theta$ and $C < \infty$ such that

$$|K(x, y)| \leq \frac{C}{\rho(x, y)} \tag{1.10}$$

for all $x \neq y$,

$$|K(x, y) - K(x', y)| \leq C\rho(x, x')^{\epsilon}\rho(x, y)^{-(1+\epsilon)} \tag{1.11}$$

for $\rho(x, x') \leq \frac{1}{2A}\rho(x, y)$,

$$|K(x, y) - K(x, y')| \leq C\rho(y, y')^{\epsilon}\rho(x, y)^{-(1+\epsilon)} \tag{1.12}$$

for $\rho(y, y') \leq \frac{1}{2A}\rho(x, y)$.

Definition 1.9 ([CW1]). A continuous linear operator $T : \dot{C}_0^{\eta}(X) \to (\dot{C}_0^{\eta}(X))'$, is said to be a Calderón-Zygmund singular integral operator if T is associated to a Calderón-Zygmund kernel K such that

$$\langle Tf, g\rangle = \int\int K(x, y)f(y)g(x)d\mu(y)d\mu(x) \tag{1.13}$$

for all f and $g \in \dot{C}_0^{\eta}(X)$ with $\text{supp}(f) \cap \text{supp}(g) = \emptyset$.

We let $CZK(\epsilon)$ denote the collection of Calderón-Zygmund singular integral operators T satisfying (1.10), (1.11) and (1.12).

Using Theorem 1.7, one can show the following result:

Theorem 1.10 ([CW1]). *Any Calderón-Zygmund singular integral operator which is bounded on $L^2(X)$ is also bounded on $L^p(X)$ for $1 < p < \infty$, and is of weak type* $(1,1)$.

We say that T is a Calderón-Zygmund operator if T is a Calderón-Zygmund singular integral and bounded on L^2.

As in the case of $\mathbb{R}^n$, we define BMO functions as follows.

Definition 1.11. Suppose $f \in L^1_{loc}(X)$. Then f belongs to $BMO(X)$ if

$$\|f\|_{BMO} = \sup_{r>0, x\in X} \frac{1}{\mu(B(x,r))} \int_{B(x,r)} |f(y) - f_B| d\mu(y) < \infty$$

where $f_B = \frac{1}{\mu(B(x,r))} \int\limits_{B(x,r)} f(y)d\mu(y)$.

As in the case of $\mathbb{R}^n$, the following result gives the endpoint estimate of the Calderón -Zygmund operators at $p = \infty$.

Theorem 1.12 ([P]). *Any Calderón-Zygmund operator maps $L^\infty(X)$ boundedly to $BMO(X)$.*

From Theorem 1.10 a question arises: Under what conditions a Calderón-Zygmund singular integral is bounded on L^2? This question was answered by the well-known $T1$ theorem of G. David and J. L. Journé, and G. David, J. L. Journé and S. Semmes in the standard case of $\mathbb{R}^n$ and in spaces of homogeneous type, respectively. The main tool they used is the Littlewood-Paley analysis which has been explained in the case of $\mathbb{R}^n$. We will describe the Littlewood-Paley analysis on spaces of homogeneous type in the next section.

1.3 Littlewood-Paley Analysis on Spaces of Homogeneous Type

We aim at estimating the L^p norm of a function f by the L^p norm of its Littlewood-Paley function $S(f)$. The proof relies on a decomposition of the identity operator suggested by R. R. Coifman.

We take a nonnegative smooth function $h(x)$ equal to 1 on $[1,9]$ and to 0 for $x \leq 0$ and $x \geq 10$. Let H_k be the operator with kernel $2^k h(2^k \rho(x,y))$. The doubling condition on μ and the construction of h imply that $C^{-1} \leq H_k(1)(x) \leq C$ for all $k \in \mathbb{Z}, x \in X$ and some constant $C > 0$. Let M_k and W_k

be the operators of pointwise multiplication by $m_k(x) = (H_k(1)(x))^{-1}$ and $w_k(x) = (H_k(m_k)(x))^{-1}$, respectively, and finally, let $S_k = M_k H_k W_k H_k M_k$. We, particularly, emphasize that this special construction will be used in Chapter 4. It is easy to check that $S_k(x, y)$, the kernel of S_k, satisfies the following conditions: There exists a constant C such that for all $k \in \mathbb{Z}$ and all $x, x', y \in X$:

(i) $S_k(x, y) = 0$ if $\rho(x, y) \geq C2^{-k}$ and $\|S_k\|_\infty \leq C2^k$;
(ii) $|S_k(x, y) - S_k(x', y)| \leq C2^{k(1+\theta)}\rho(x, x')^\theta$;
(iii) $\int_X S_k(x, y)d\mu(y) = 1$;
(iv) $S_k(x, y) = S_k(y, x)$.

From (i) and (iii), it is clear that $\lim_{k\to\infty} S_k = I$, the identity operator on $L^2(X)$, and $\lim_{k\to-\infty} S_k = 0$ in the strong operator topology on $L^2(X)$. The construction of wavelets on a space of homogeneous type (see Theorem 3.25) will be based on this approximation to the identity.

We now come to Coifman's decomposition of the identity operator. Set $D_k = S_k - S_{k-1}$ so that $I = \sum_{k=-\infty}^{\infty} D_k$ with strong convergence. The idea of Coifman is to rewrite

$$I = \sum_{k=-\infty}^{\infty} D_k \sum_{j=-\infty}^{\infty} D_j = \sum_{|k-j|\leq N} D_k D_j + \sum_{|k-j|>N} D_k D_j = T_N + R_N.$$

We prove that $(T_N)^{-1}$ exists on L^2 as N is large. More precisely, we have

Theorem 1.13. *Let T_N be the same as above. Then T_N is invertible on $L^2(X)$ when N is large enough. Moreover, $(T_N)^{-1}$, the inverse of T_N, is uniformly bounded on $L^2(X)$ when $N \geq N_0$.*

By the almost orthogonal argument (see the proof of Lemma 3.7 below in a more general case), $D_k D_j(x, y)$, the kernel of $D_k D_j$, satisfies the following estimate

$$|D_k D_j(x, y)| \leq C2^{-|j-k|\theta}2^{(k\wedge j)}\chi(\{\rho(x, y) \leq C2^{-(k\wedge j)}\}) \tag{1.14}$$

where $k \wedge j = \min(k, j)$ and χ is the indicator function.

The estimate in (1.14) allows one to apply the famous Cotlar-Stein Lemma ([MC]).

Lemma 1.14. *Suppose that $\{T_j\}$ is a finite collection of bounded operators on some Hilbert space H satisfying the following estimates*

$$\|T_i T_j^*\| \leq [\gamma(i - j)]^2,$$

and

$$\|T_i^* T_j\| \leq [\gamma(i - j)]^2,$$

where $A = \sum_{j=-\infty}^{\infty} \gamma(j) < \infty$ *Then* $T = \sum_j T_j$ *is bounded on* H *(this being obvious) with* $\|T\| \le A$.

The estimate in (1.14) together with the Cotlar-Stein Lemma 1.14 implies that the operator R_N is bounded on L^2 with an operator norm not larger than $C2^{-N\delta}$ for some fixed constant $C > 0$ and $\delta > 0$. This yields that T_N converges to the identity, as an operator on L^2, when $N \to \infty, T_N$ is invertible on L^2 as N is large, and $(T_N)^{-1}$, the inverse of T_N, is uniformly bounded on $L^2(X)$ for large N. This ends the proof of Theorem 1.13.

One observes that the operator R_N is a Calderón-Zygmund operator with an operator norm at most $C2^{-N\delta}$. This implies that T_N is bounded on $L^p, 1 < p < \infty$, and $(T_N)^{-1}$ is also uniformly bounded on $L^p, 1 < p < \infty$, for large N.

We now enter the proof of the Littlewood-Paley estimate

$$C^{-1}\|f\|_2 \le C\left\|\left(\sum_k |D_k f|^2\right)^{\frac{1}{2}}\right\|_2 \le C\|f\|_2.$$

We first prove the right-hand side by a duality argument which runs as follows

$$\begin{aligned}
&\sup\left\{\left(\sum_k \|D_k f\|_2^2\right)^{\frac{1}{2}} : f \in L^2, \|f\|_2 \le 1\right\}\\
&= \sup\left\{\left|\sum_k \langle D_k f, g_k\rangle\right| : f \in L^2, \|f\|_2 \le 1, \sum_k \|g_k\|_2^2 \le 1\right\}\\
&= \sup\left\{\left\|\sum_k D_k^* g_k\right\|_2 : \sum_k \|g_k\|_2^2 \le 1\right\}.
\end{aligned}$$

But

$$\begin{aligned}
\left\|\sum_k D_k^* g_k\right\|_2^2 &= \left\langle \sum_k D_k^* g_k, \sum_i D_i^* g_i \right\rangle = \sum_k \sum_i \langle D_i D_k^* g_k, g_i\rangle\\
&\le \sum_k \sum_i \|D_i D_k^*\|_{2,2} \|g_k\|_2 \|g_i\|_2\\
&\le C \sum_k \|g_k\|_2^2
\end{aligned}$$

where the last inequality follows from the estimate in (1.14). This implies

$$\sum_k \|D_k f\|_2^2 \le C\|f\|_2^2.$$

We now turn to the left-hand side. Again, by a duality argument, we have

$$\begin{aligned}\|T_N f\|_2 &= \sup\Big\{|\langle T_N f, g\rangle| : \|g\|_2 \le 1\Big\}\\ &\le \sup\Big\{(2N+1)\big(\sum_k \|D_k f\|_2^2\big)^{\frac{1}{2}}\big(\sum_k \|D_k g\|_2^2\big)^{\frac{1}{2}} : \|g\|_2 \le 1\Big\}\\ &\le C(2N+1)\big(\sum_k \|D_k f\|_2^2\big)^{\frac{1}{2}}\end{aligned}$$

where the last inequality follows from the above estimate.

The integer N will be viewed as a large constant in what follows. This constant will be incorporated inside the generic constant C. In other words the reader should not raise the issue of considering limits as N tends to ∞. The estimates above together with Theorem 1.13 yield the following Littlewood-Paley estimate on L^2.

$$\begin{aligned}&C^{-1}\|f\|_2\\ &\quad = C^{-1}\|(T_N)^{-1}T_N f\|_2 \le \|T_N(f)\|_2 \le C\Big\|\Big(\sum_k |D_k f|^2\Big)^{\frac{1}{2}}\Big\|_2 \le C\|f\|_2.\end{aligned}$$

By a routine argument, for example considering $S(f) = \Big(\sum\limits_k |D_k f|^2\Big)^{\frac{1}{2}}$ as a vector-valued Calderón-Zygmund operator, we obtain

$$\Big\|\Big(\sum_k |D_k f|^2\Big)^{\frac{1}{2}}\Big\|_p \le C\|f\|_p$$

where $1 < p < \infty$.

The above estimate together with the identity on L^2

$$f = (T_N)^{-1}\sum_k D_k^N D_k f$$

provides the Littlewood-Paley estimates on $L^p, 1 < p < \infty$,

$$C^{-1}\|f\|_p \le \Big\|\Big(\sum_k |D_k f|^2\Big)^{\frac{1}{2}}\Big\|_p \le C\|f\|_p.$$

This is the Littlewood-Paley analysis of G. David, J. L. Journé and S. Semmes on spaces of homogeneous type. We are now ready to prove the $T1$ theorem of G. David, J. L. Journé and S. Semmes, which will be given in the next section.

1.4 The $T1$ Theorem on Spaces of Homogeneous Type

In this section we prove the $T1$ theorem of G. David, J. L. Journé and S. Semmes on spaces of homogeneous type.

We suppose that $\mu(X) = \infty$ and $\mu(\{x\}) = 0$ for all $x \in X$. We give necessary and sufficient conditions for $L^2(X)$ boundedness of a Calderón-Zygmund singular integral operator T. This statement is the so-called $T1$ theorem. To formulate the $T1$ theorem, we need the definition of weak boundedness. For $\delta \in (0, \theta], x_0 \in X$ and $r > 0$, we define $A(\delta, x_0, r)$ to be the set of all $\phi \in \dot{C}_0^{\delta}(X)$ supported in $B(x_0, r)$ satisfying $\|\phi\|_\infty \leq 1$ and $\|\phi\|_\delta \leq r^{-\delta}$.

Definition 1.15. An operator T is weakly bounded if there exist $0 < \delta \leq \theta$ and $C < \infty$ such that for all $x_0 \in X, r > 0$, and $\phi, \psi \in A(\delta, x_0, r)$,

$$|\langle T\phi, \psi \rangle| \leq C\mu(B(x_0, r)). \tag{1.15}$$

If T is weakly bounded, we write $T \in WBP$.

It is easy to see that weak boundedness is obviously implied by L^2 boundedness. Calderón-Zygmund singular integral operator whose kernel is antisymmetrical kernel, that is $K(x, y) = -K(y, x)$, has the weak boundedness property.

We now define $T(1)$. The difficulty is that 1 is not a function in $\dot{C}_0^{\eta}(X)$. This will lead to the fact that $T(1)$ is not a distribution in $(\dot{C}_0^{\eta})'$, but is a distribution modulo constant functions. The definition is based on the following lemma ([MC]).

Lemma 1.16. *Let S be a distribution in $(\dot{C}_0^{\eta})'$. Suppose that there exists $R > 0$ such that the restriction of S to the open set $\{x \in X : \rho(x, x_0) > R\}$, where x_0 is a fixed point in X, is a continuous function and such that $S(x) = O(\rho(x, x_0)^{-1-\gamma})$ as $\rho(x, x_0) \to \infty$. If $\gamma > 0$, then the integral*

$$\int_X S(x) d\mu(x) = \langle S, 1 \rangle \tag{1.16}$$

converges.

We first write $1 = \phi_1(x) + \phi_2(x)$, where $\phi_1 \in \dot{C}_0^{\eta}(X)$ for some $\eta > 0$ and $\phi_1(x) = 1$ for $\rho(x, x_0) \leq R$. Then $\langle S, 1 \rangle$ is defined by $\langle S, \phi_1 \rangle + \langle S, \phi_2 \rangle = \langle S, \phi_1 \rangle + \int S(x)\phi_2(x) d\mu(x)$ since the integral converges absolutely. It is easy to check that $\langle S, 1 \rangle$ is independent of the decomposition.

Before defining $T1$, we use the following definition.

Definition 1.17. Let $\dot{C}_{0,0}(X) \subset \dot{C}_0(X)$ be defined by $\int\limits_X f(x) d\mu(x) = 0$.

If $f \in \dot{C}_{0,0}^{\eta}(X)$, we define $\langle T1, f \rangle = \langle 1, T^* f \rangle$. In fact, if the support of f is contained in $\{x \in X : \rho(x, x_0) \leq R\}$, then

$$T^*(f)(x) = \int [K(y,x) - K(x_0,x)]f(y)d\mu(y) = O\big(\rho(x,x_0)^{-1-\epsilon}\big)$$

for $\rho(x,x_0) > R$ and $\epsilon > 0$.

Now $T1$ is a continuous linear form on $\dot{C}^\eta_{0,0}(X) \subset \dot{C}^\eta_0(X)$. We extend $T1$ to a distribution $S \in (\dot{C}^\eta_0)'$ as follows: Let $\phi \in \dot{C}^\eta_0(X)$ be a function with $\int \phi(x)d\mu(x) = 1$, then for all $f \in \dot{C}^\eta_0(X), f$ can be written uniquely as $f = \lambda\phi + g$, where $\lambda = \int f(x)d\mu(x)$ and $g \in \dot{C}^\eta_{0,0}(X)$. Now choose $\langle S,\phi\rangle$ arbitrarily and put $\langle S,f\rangle = \lambda\langle S,\phi\rangle + \langle 1,g\rangle$. So $T1 = S$ on $\dot{C}^\eta_{0,0}(X)$, and is a distribution modulo the constant. T^*1 can be defined by a similar way.

We now state and prove the following $T1$ theorem ([DJS]).

Theorem 1.18. *Let T be a Calderón-Zygmund singular integral operator. Then a necessary and sufficient condition for the extension of T as a continuous linear operator on $L^2(X)$ is that the following properties are all satisfied: (a) $T1 \in BMO$, (b) $T^*1 \in BMO$, (c) T is weakly bounded.*

Indeed, by Theorem 1.12, we only need to prove the sufficient condition. Before proving Theorem 1.18, we first outline the general philosophy of the proof of the $T1$ theorem when $X = \mathbb{R}^n$. The $T1$ theorem on $\mathbb{R}^n$ was proved in two steps. In the first step, one considered the case $T(1) = T^*(1) = 0$. For this step a clever approximation of the identity operator is used. To be precise, let $\psi \in C_0^\infty(\mathbb{R}^n)$ be radial and satisfy

$$\int_0^\infty |\widehat{\psi}(t\xi)|^2 \frac{dt}{t} = 1$$

for all $\xi \neq 0$. Let $\psi_t(x) = \frac{1}{t^n}\psi(\frac{x}{t})$ and Q_t be the operator of convolution with ψ_t. The operator

$$U_m = \int_{\frac{1}{m}}^m Q_t^2 \frac{dt}{t}$$

converges strongly to the identity, as an operator on $L^2(\mathbb{R}^n)$, when $m \to \infty$. Since U_m is continuous on $C_0^\infty(\mathbb{R}^n)$ and on its dual, so U_mTU_n is well defined from $C_0^\infty(\mathbb{R}^n)$ into its dual. For all m, n, T will be bounded on L^2 if and only if U_mTU_n is bounded on L^2 with a norm that is independent of m and n. To show that U_mTU_n is bounded on L^2 uniformly for m and n, the regularity conditions of the kernel, the weak boundedness property of T, the conditions $T(1) = T^*(1) = 0$, and the Littlewood-Paley theory are playing a key role. This Littlewood-Paley theory replaces the Cotlar-Stein almost orthogonality lemma which was used in the first proof. In the second step we know that $T(1) = \beta$ and $T^*(1) = \gamma$ lie in BMO. Then David and Journé used the so-called para-product operators Π_β and $(\Pi_\gamma)^*$ which are Calderón-Zygmund operators whenever β and γ belong to BMO. The L^2 continuity of these para-products is equivalent to the characterization of BMO through Carleson measures. Moreover, $\Pi_\beta(1) = \beta$ and $(\Pi_\beta)^*(1) = 0$. Thus, $T - \Pi_\beta - (\Pi_\gamma)^*$ is bounded on L^2 by the first step, and hence, T is bounded on L^2.

We now return to spaces of homogeneous type and to the proof of the first step of Theorem 1.18. We assume that $T(1) = T^*(1) = 0$. The proof relies on the Littlewood-Paley estimate on L^2 and estimates which will be proved later in this book. Roughly speaking if T satisfies all conditions of Theorem 1.18 and $T(1) = T^*(1) = 0$, then a similar estimate as in (1.14) still holds. Indeed the matrix Ω whose entries are the operator norms $\omega(j,k) = \|D_jTD_k\|_{2,2}$ is bounded on l^2 (see Lemma 4.23 below). Observe, by the Littlewood -Paley estimate on L^2, that for all $f, g \in \dot{C}_0^\eta$,

$$\begin{aligned}
|\langle T_NTT_N(f), g\rangle| &= |\sum_j\sum_k \langle D_j^N D_jTD_kD_k^N(f), g\rangle| \\
&\leq C\sum_j\sum_k \|D_jTD_k\|_{2,2}\|D_k^N f\|_2\|(D_j^N)^* g\|_2 \\
&\leq C(2N+1)^2\Big(\sum_k \|D_kf\|_2^2\Big)^{\frac{1}{2}}\Big(\sum_k \|D_kg\|_2^2\Big)^{\frac{1}{2}} \\
&\leq C(2N+1)^2\|f\|_2\|g\|_2
\end{aligned}$$

where the estimate given in Lemma 4.23 applies to the estimate on $\|D_jTD_k\|_{2,2}$.

To finish the proof of the first step of Theorem 1.18 we now use Theorem 1.13 together with the following lemma. The exponent θ is defined in (1.7).

Lemma 1.19. *If $0 < s < \eta$ the series $T_N = \sum_j D_jD_j^N$ converges strongly on $\dot{C}_0^s$. If $0 < 2s < \theta, T_N$ is bounded on $\dot{C}_0^s$, and the operator norm of $I - T_N$ on $\dot{C}_0^s$ tends to 0 as $N \to \infty$.*

The space $\dot{C}_0^s$ is defined in Theorem 1.3. Let us use this lemma to prove that T extends to a bounded operator on L^2. First, we can use the size estimate of the kernel of T to extend T to a continuous linear operator from $\dot{C}_0^s \cap L^2$ into $(\dot{C}_0^s)'$. To see this, let $f \in \dot{C}_0^s \cap L^2$ and $g \in \dot{C}_0^s$, and choose $\theta(x) \in \dot{C}_0^s$ with $\theta(x) = 1$ on a neighborhood of the support of g, then we can define $\langle g, Tf\rangle = \langle g, T(\theta f)\rangle + \langle g, T(1-\theta)f\rangle$. The first term makes sense since $\theta f \in \dot{C}_0^s$, while the second term can be defined by the size estimate of the kernel of T and the assumption $f \in L^2$. It is clear that this definition doesn't depend on θ. In view of the claim, T_N is defined and bounded on $\dot{C}_0^s \cap L^2$, with norm $\|\cdot\|_s + \|\cdot\|_2$, and it is invertible on $\dot{C}_0^s \cap L^2$ if N is large enough. Notice that the series $\sum_j D_jD_j^N$ converges strongly on L^2 since $\sup_{m,n}\Big\|\sum_{j=-n}^m D_jD_j^N\Big\|_{2,2} < \infty$ and $\sum_j D_jD_j^N f$ converges in L^2 if $f \in \dot{C}_0^\eta$. Thus $\sum_j D_jD_j^N$ converges strongly on $\dot{C}_0^s \cap L^2$, by the lemma.

Let $f_1 \in \dot{C}_0^s$ have compact support and let $f_2 = T_N^{-1} f_1 \in \dot{C}_0^s \cap L^2$, so that

$$\lim_{n\to\infty} \sum_{j=-n}^{n} D_j D_j^N f_2 = f_1$$

in $\dot{C}_0^s \cap L^2$. If $g \in L^2$ with compact support, then

$$\begin{aligned}\left|\left\langle \sum_{j=-n}^{n} D_j D_j^N g,\ Tf_1 \right\rangle\right| &= \left| \lim_{m\to\infty} \left\langle \sum_{j=-n}^{n} D_j D_j^N g,\ T \sum_{j=-m}^{m} D_j D_j^N f_2 \right\rangle\right| \\ &\le C\|g\|_2 \|f_2\|_2 \le C\|g\|_2\|f_1\|_2.\end{aligned}$$

A similar argument allows one to show that if $g_1 \in \dot{C}_0^s$ has compact support, then $|\langle g_1, Tf_1\rangle| \le C\|g_1\|_2\|f_1\|_2$. Thus T extends to a bounded operator on L^2.

We now return to the proof of Lemma 1.19. The norm in $\dot{C}_0^s$ will be simply denoted by $\|\cdot\|_s$ and cannot be confused with a norm in a Lebesgue space. To do this, we need the following preliminary facts:

(a) if $0 < \beta < \gamma$, then $\|f\|_\beta \le \|f\|_\infty^{\frac{\gamma-\beta}{\gamma}} \|f\|_\gamma^{\frac{\beta}{\gamma}}$;

(b) $\|D_k\|_\infty + \|S_k\|_\infty \le C$ and $\|D_k f\|_\theta \le C2^{k\theta}\|f\|_\infty$;

(c) if $0 < s < \beta, g_k \in L^\infty \cap \dot{C}_0^\beta, \|g_k\|_\infty \le 2^{-ks}$ and $\|g_k\|_\beta \le 2^{k(\beta-s)}$, then $\left\|\sum_k g_k\right\|_s \le C(s,\beta)$, where the series converges locally in the norm $\|\cdot\|_\eta$ for $0<\eta<s$, but not necessarily in $\dot{C}_0^s$.

It is easy to verify (a) and (b) follow from the definitions of S_k and D_k. For (c), let $x, y \in X$ be given and choose k_0 such that $2^{-k_0} \le \rho(x,y) \le 2^{-k_0+1}$. Then

$$\begin{aligned}\left|\sum_k g_k(x) - \sum_k g_k(y)\right| &\le \sum_{k\ge k_0} |g_k(x) - g_k(y)| + \sum_{k<k_0} |g_k(x) - g_k(y)| \\ &\le \sum_{k\ge k_0} 2\|g_k\|_\infty + \sum_{k<k_0} \rho(x,y)^\beta \|g_k\|_\beta \\ &\le C2^{-k_0 s} + C2^{k_0(\beta-s)}\rho(x,y)^\beta \\ &\le C\rho(x,y)^s.\end{aligned}$$

Suppose that $g \in \dot{C}_0^s$. Since $D_k(1) = 0$,

$$\begin{aligned}|D_k(g)(x)| &= \left|\int D_k(x,y) g(y) d\mu(y)\right| \\ &= \left|\int D_k(x,y)[g(y) - g(x)] d\mu(y)\right| \le C2^{-ks}\|g\|_s;\end{aligned}$$

applying (b) and (c) to $g_k = D_k D_k^N g$ yields $\left\| \sum_k D_k D_k^N g \right\|_s \le C_N \|g\|_s$. If $g \in \dot{C}_0^r$ for some $r > s$, then $\sum_k D_k D_k^N g$ converges uniformly and in $\dot{C}_0^s$ norm. Thus, $\sum_k D_k D_k^N$ is defined and bounded on $\dot{C}_0^s$. To show the operator norm of $I - T_N$ on $\dot{C}_0^s$ tends to 0 as $N \to \infty$, we write $(I - T_N)g = \sum_k D_k(I - S_{k+N})g + \sum_k D_k S_{k-N-1} g$. Since the kernel of S_k is 0 when $\rho(x,y) \ge C2^{-k}$ and $(I - S_{N+k})(1) = 0$, so $\|(I - S_{N+k})g\|_\infty \le C2^{-(N+k)s}\|g\|_s$. If $g_k = D_k(I - S_{k+N})g$, then (b) and (c) imply that $\left\| \sum_k g_k \right\|_s \le C2^{-Ns}\|g\|_s$.

Let $g_k = D_k S_{k-N-1} g$, and let $L_{k,N}(x,y)$ denote the kernel of $D_k S_{k-N-1}$. Since $L_{k,N}(1) = 0$, $L_{k,N}(x,y) = 0$ if $\rho(x,y) \ge C2^{-(k-N)}$, and $|L_{k,N}(x,y)| \le C2^{(k-N)(1+\theta)-k\theta}$. Thus,

$$\begin{aligned} \|g_k\|_\infty &\le 2^{-(k-N)s} 2^{(k-N)(1+\theta)-k\theta-(k-N)} \|g\|_s \\ &\le C2^{-N\theta-(k-N)s}\|g\|_s. \end{aligned}$$

If $\rho(x,x') \le C2^{-(k-N)}$, then

$$|g_k(x) - g_k(x')| \le C2^{k\theta-(k-N)s}\rho(x,x')^\theta \|g\|_s.$$

When $\rho(x,x') \ge C2^{-(k-N)}$,

$$|g_k(x) - g_k(x')| \le 2\|g_k\|_\infty \le C2^{k\theta-(k-N)s}\rho(x,x')^\theta \|g\|_s.$$

Thus,

$$\|g_k\|_\theta \le C2^{k\theta-(k-N)s}\|g\|_s.$$

Using (a), we get that

$$\|g_k\|_\beta \le C2^{k\beta-N(\theta-\beta)-(k-N)s}\|g\|_s \le C2^{-N(\theta-2\beta)+(k-N)(\beta-s)}\|g\|_s.$$

Applying (c) to $\widetilde{g}_k = 2^{N(\theta-2\beta)} g_{k+N}$, it follows that

$$\left\| \sum_k g_k \right\|_s \le C2^{-N(\theta-2\beta)}\|g\|_s$$

if $s < \beta < \theta$.

If $2s < \theta$, we can choose β so that $2^{-N(\theta-2\beta)}$ tends to 0 as $N \to \infty$, which implies Lemma 1.19.

We turn to the second step of the proof of Theorem 1.18. It consists in building, for each $b \in BMO$, a Calderón-Zygmund operator Π_b, the paraproduct operator, such that $\Pi_b(1) = b$ and $(\Pi_b)^*(1) = 0$. For this purpose, define for $b \in BMO$,

$$\Pi_b(f)(x) = \sum_k D_k^N\{D_k(T_N)^{-1}(b)(\cdot)S_k(f)(\cdot)\}(x).$$

To see that the operator Π_b is well defined, we show that T_N is invertible also on BMO for large N. Observe that $I - T_N = \sum_k D_k(I - D_k^N) = \sum_k D_k(I - S_{k+N}) - \sum_k D_k S_{k-N-1}$. It is not difficult to check that the kernels of the operators $\sum_k D_k(I - S_{k+N})$ and $\sum_k D_k S_{k-N-1}$ satisfy the estimate in (1.10) with the constant $C2^{-\epsilon N}$, which follows from the facts that $(I - S_{k+N})(1) = 0$ and $D_k(1) = 0$ together with the smoothness in x of the kernel of S_{k-N-1}. The kernel of $I - T_N$ also satisfies the estimate in (1.11) with a constant not depending on N. See more details of these estimates in Section 3 of Chapter 3. The geometric mean of these two estimates on $I - T_N$ implies that the kernel of $I - T_N$ satisfies the estimate in (1.11) with ϵ replaced by $\frac{\epsilon}{2}$ and a constant $C2^{-N\frac{\epsilon}{2}}$. These estimates on $I - T_N$ together with the facts that $(I - T_N)(1) = 0$ in BMO and $I - T_N$ has L^2-operator norm at most $C2^{-N\delta}, \delta > 0$ imply that $I - T_N$ has a small BMO -operator norm when N is large. Therefore T_N is invertible on BMO if N is large.

To show that Π_b is bounded on L^2 for $b \in BMO$, it suffices to prove that

$$\sum_k |\langle D_k^N\{D_k(T_N)^{-1}(b)(\cdot)S_k(f), g\rangle| \leq C\|f\|_2\|g\|_2$$

for $f, g \in L^2$. Using two times Cauchy-Schwarz inequality we obtain

$$\begin{aligned}
&\sum_k |\langle D_k^N\{D_k(T_N)^{-1}(b)(\cdot)S_k(f), g\rangle| \\
&\quad \leq \sum_k \|(D_k^N)^*(g)\|_2 \|D_k(T_N)^{-1}(b)(\cdot)S_k(f)\|_2 \\
&\quad \leq \left\{\sum_k \|(D_k^N)^*(g)\|_2^2\right\}^{\frac{1}{2}} \left\{\sum_k \|D_k(T_N)^{-1}(b)(\cdot)S_k(f)\|_2^2\right\}^{\frac{1}{2}}.
\end{aligned}$$

By the Littlewood-Paley estimate on L^2, the first factor is at most $C\|g\|_2$. This estimate also implies that $d\mu(x,t) = \sum_k |D_k(T_N)^{-1}(b)|^2 d\mu(x) d\delta_k(t)$ is a Carleson measure on $X \times (0, \infty)$, where $\delta_k(t)$ is the Dirac measure at 2^{-k}. We define a function $F(x,t)$ by $F(x,t) = S_k(f)(x)$ if $t = 2^{-k}$ and 0 otherwise. Then $F(x,t)$ has a nontangential maximal function dominated by the Hardy-Littlewood maximal function of f. From Carleson's inequality, it follows that

$$\begin{aligned}
\sum_k \|D_k(T_N)^{-1}(b)(\cdot)S_k(f)\|_2^2 &= \iint |F(x,t)|^2 d\mu(x,t) \\
&\leq C\|(T_N)^{-1}(b)\|_{BMO}^2 \|f\|_2^2 \leq C\|b\|_{BMO}^2 \|f\|_2^2.
\end{aligned}$$

See [MC] for more details about the Carleson measure and Carleson's inequality.

Moreover, $\Pi_b(1) = \sum_k D_k^N\{D_k(T_N)^{-1}(b)\}(x) = b(x)$ and $(\Pi_b)^*(1) = 0$.

Define $U = T - \Pi_{T(1)} - (\Pi_{T^*(1)})^*$. Then U satisfies the hypotheses of the theorem, moreover, $U(1) = U^*(1) = 0$. So U is bounded on $L^2(X)$ by the first step of the proof of Theorem 1.18, and hence T is bounded on $L^2(X)$. This completes the proof of Theorem 1.18.

As it is well known in the standard case of $\mathbb{R}^n$, the construction of wavelet cannot be obtained from classical Littlewood-Paley expansions. Multiresolution analysis is seminal in the construction of wavelet bases. This also happens for spaces of homogeneous type. Indeed the Littlewood-Paley analysis developed by G. David, J. L. Journé and S. Semmes cannot provide wavelet expansions on spaces of homogeneous type. In the next chapter, we will prove a new "$T1$ theorem", namely the boundedness of Calderón-Zygmund operators on a *wavelet space*. This new "$T1$ theorem" will be a main tool to provide a wavelet analysis.

Chapter 2
The Boundedness of Calderón-Zygmund Operators on Wavelet Spaces

We first define test functions and wavelet spaces on spaces of homogeneous type. Then we prove the main result of this chapter, namely that Calderón-Zygmund operators whose kernels satisfy an additional smoothness condition are bounded on *wavelet spaces*. This result will be a crucial tool to provide wavelet expansions of functions and distributions on spaces of homogeneous type in the next chapter.

We first introduce test functions on spaces of homogeneous type.

Definition 2.1. Fix $0 < \gamma, \beta < \theta$. A function f defined on X is said to be a test function of type (x_0, r, β, γ), $x_0 \in X$, and $r > 0$, if f satisfies the following conditions:

(i) $|f(x)| \leq C \frac{r^\gamma}{(r+\rho(x,x_0))^{1+\gamma}}$;

(ii) $|f(x) - f(y)| \leq C\Big(\frac{\rho(x,y)}{r+\rho(x,x_0)}\Big)^\beta \frac{r^\gamma}{(r+\rho(x,x_0))^{1+\gamma}}$ for all $x, y \in X$ with $\rho(x,y) \leq \frac{1}{2A}(r + \rho(x, x_0))$.

Such functions exist and the reader will find a recipe two lines after Definition 1.2. If f is a test function of type (x_0, r, β, γ), we write $f \in \mathcal{M}(x_0, r, \beta, \gamma)$, and the norm of f in $\mathcal{M}(x_0, r, \beta, \gamma)$ is defined by

$$\|f\|_{\mathcal{M}(x_0,r,\beta,\gamma)} = \inf\{C : (i) \text{ and } (ii) \text{ hold}\}.$$

One should observe that if $f \in \mathcal{M}(x_0, r, \beta, \gamma)$, then

$$\|f\|_1 \approx \|f\|_{\mathcal{M}(x_0,r,\beta,\gamma)}.$$

We say that a function f is a *scaling function* if $f \in \mathcal{M}(x_0, r, \beta, \gamma)$ and $\int f(x)d\mu(x) = 1$.

Now fix $x_0 \in X$ and denote $\mathcal{M}(\beta, \gamma) = \mathcal{M}(x_0, 1, \beta, \gamma)$. It is easy to see that $\mathcal{M}(x_1, r, \beta, \gamma) = \mathcal{M}(\beta, \gamma)$ with equivalent norms for all $x_1 \in X$ and $r > 0$. Furthermore, it is also easy to check that $\mathcal{M}(\beta, \gamma)$ is a Banach space with respect to the norm in $\mathcal{M}(\beta, \gamma)$.

D. Deng and Y. Han, *Harmonic Analysis on Spaces of Homogeneous Type*, Lecture Notes in Mathematics 1966,

Definition 2.2. A function f defined on X is said to be a wavelet of type (x_0, r, β, γ) if $f \in \mathcal{M}(x_0, r, \beta, \gamma)$ and $\int f(x)d\mu(x) = 0$. We denote this by $f \in \mathcal{M}_0(x_0, r, \beta, \gamma)$.

These wavelets are named *molecules* by Guido Weiss. A compactly supported molecule is an *atom*. Atomic decompositions preluded wavelet analysis, as indicated in the Introduction. Moreover Caderón-Zygmund operators T satisfying $T(1) = T^*(1) = 0$ have the remarkable property map a molecule into a molecule. We use the notation $\mathcal{M}_0(\beta, \gamma)$, when the dependence in x_0 and r can be forgotten, as a space of wavelets with regularity (β, γ).

To study the boundedness of Calderón-Zygmund singular integral operators on a *wavelet space*, we define the following "strong" weak boundedness property.

Definition 2.3. An operator T defined by a distributional kernel K, is said to have the "strong weak boundedness property" if there exist $\eta > 0$ and $C < \infty$ such that

$$|\langle K, f\rangle| \leq Cr \tag{2.1}$$

for all $f \in C_0^\eta(X \times X)$ with $\mathrm{supp}(f) \subseteq B(x_1, r) \times B(y_1, r), x_1$ and $y_1 \in X, \|f\|_\infty \leq 1, \|f(\cdot, y)\|_\eta \leq r^{-\eta}$, and $\|f(x, \cdot)\|_\eta \leq r^{-\eta}$ for all x and $y \in X$.

If T has the "strong weak boundedness property", we write $T \in SWBP$.

Note that if ψ and ϕ are functions satisfying the conditions in Definition 1.15, then $f(x, y) = \psi(x) \times \phi(y)$ satisfies the conditions in Definition 2.3, and hence $|\langle T\psi, \phi\rangle| = |\langle K, f\rangle| \leq Cr$ if T has the "strong weak boundedness property". This means that the strong weak boundedness property implies the weak boundedness property. However, in the standard situation of $\mathbb{R}^n$, the weak boundedness property implies the strong one. Indeed any smooth function $f(x, y), x \in B, y \in B$, supported by $B \times B$ can be written, by a double Fourier series expansion, as $\sum \alpha_j f_j(x) g_j(y)$ with $\sum |\alpha_j| < \infty, \|f_j\|_{C_0^\beta} \leq 1, \|g_j\|_{C_0^\beta} \leq 1$.

If $T \in CZK(\epsilon)$, we say that $T^*(1) = 0$ if $\int T(f)(x)dx = 0$ for all $f \in \mathcal{M}_0(\beta, \gamma)$. Similarly, $T(1) = 0$ if $\int T^*(f)(x)dx = 0$ for all $f \in \mathcal{M}_0(\beta, \gamma)$.

The main result in this chapter is the following theorem.

Theorem 2.4. *Suppose that $T \in CZK(\epsilon) \cap SWBP$, and $T(1) = T^*(1) = 0$. Suppose further that $K(x, y)$, the kernel of T, satisfies the following condition:*

$$\begin{aligned} &|K(x, y) - K(x', y) - K(x, y') + K(x', y')| \\ &\leq C\rho(x, x')^\epsilon \rho(y, y')^\epsilon \rho(x, y)^{-(1+2\epsilon)} \end{aligned} \tag{2.2}$$

for $\rho(x, x'), \rho(y, y') \leq \frac{1}{2A}\rho(x, y)$. Then there exists a constant C such that for each wavelet $f \in \mathcal{M}_0(x_0, r, \beta, \gamma)$ with $x_0 \in X, r > 0$ and $0 < \beta, \gamma < \epsilon$, $Tf \in \mathcal{M}_0(x_0, r, \beta, \gamma)$. Moreover

$$\|T(f)\|_{\mathcal{M}(x_0,r,\beta,\gamma)} \leq C\|T\|\|f\|_{\mathcal{M}(x_0,r,\beta,\gamma)} \tag{2.3}$$

where $\|T\|$ denote the smallest constant in the "strong weak boundedness property" and in the estimates of the kernel of T.

Before proving Theorem 2.4, we observe that this theorem will provide wavelet expansions which, as in the standard case of $\mathbb{R}^n$, will be the building blocks of most functional spaces.

To prove Theorem 2.4, we first need the following lemma.

Lemma 2.5. *Suppose that T is a continuous linear operator from $\dot{C}_0^\eta$ to $(\dot{C}_0^\eta)'$ satisfying $T \in CZK(\epsilon) \cap SWBP$ with $\eta < \epsilon$, and $T(1) = 0$. Then there exists a constant C such that*

$$\|T\phi\|_\infty \leq C \tag{2.4}$$

whenever there exist $x_0 \in X$ and $r > 0$ such that $\mathrm{supp}\phi \subseteq B(x_0, r)$ *with $\|\phi\|_\infty \leq 1$ and $\|\phi\|_\eta \leq r^{-\eta}$.*

Proof. We follow the idea of the proof in [M1]. Fix a function $\theta \in C^\infty(\mathbb{R})$ with the following properties: $\theta(x) = 1$ for $|x| \leq 1$ and $\theta(x) = 0$ for $|x| > 2$. Let $\chi_0(x) = \theta(\frac{\rho(x,x_0)}{2r})$ and $\chi_1 = 1 - \chi_0$. Then $\phi = \phi\chi_0$ and for all $\psi \in C_0^\eta(X)$,

$$\begin{aligned}
\langle T\phi, \psi\rangle &= \langle K(x,y), \phi(y)\psi(x)\rangle = \langle K(x,y), \chi_0(y)\phi(y)\psi(x)\rangle \\
&= \langle K(x,y), \chi_0(y)[\phi(y) - \phi(x)]\psi(x)\rangle + \langle K(x,y), \chi_0(y)\phi(x)\psi(x)\rangle \\
&:= p + q
\end{aligned}$$

where $K(x,y)$ is the distribution kernel of T.

To estimate p, let $\lambda_\delta(x,y) = \theta(\frac{\rho(x,y)}{\delta})$. Then

$$\begin{aligned}
p &= \langle K(x,y), (1 - \lambda_\delta(x,y))\chi_0(y)[\phi(y) - \phi(x)]\psi(x)\rangle \\
&\quad + \langle K(x,y), \lambda_\delta(x,y)\chi_0(y)[\phi(y) - \phi(x)]\psi(x)\rangle \\
&:= p_{1,\delta} + p_{2,\delta}.
\end{aligned} \tag{2.5}$$

Since K is locally integrable on $\Omega = \{(x,y) \in X \times X : x \neq y\}$, the first term on the right hand side of (2.5) satisfies

$$\begin{aligned}
|p_{1,\delta}| &= \left| \int_\Omega K(x,y)(1 - \lambda_\delta(x,y))\chi_0(y)[\phi(y) - \phi(x)]\psi(x)d\mu(x)d\mu(y) \right| \\
&\leq C \int_X \int_X |K(x,y)\chi_0(y)[\phi(y) - \phi(x)]\psi(x)|d\mu(x)d\mu(y) \\
&\leq C \int_X |\psi(x)|d\mu(x) = C\|\psi\|_1.
\end{aligned}$$

Thus it remains to show that $\lim\limits_{\delta\to 0} p_{2,\delta} = 0$, i.e.,

$$\lim_{\delta\to 0}\langle K(x,y),\lambda_\delta(x,y)\chi_0(y)[\phi(y)-\phi(x)]\psi(x)\rangle=0, \tag{2.6}$$

and it is here that we use the "strong" weak boundedness property of T :

$$|\langle K,f\rangle|\le Cr \tag{2.7}$$

for all $f\in C_0^\eta(X\times X)$ satisfying supp$f\subseteq B(x_0,r)\times B(y_0,r), \|f\|_\infty\le 1, \|f(\cdot,y)\|_\eta\le r^{-\eta}$ and $\|f(x,\cdot)\|_\eta\le r^{-\eta}$ for all $x,y\in X$.

To show (2.6), let $\{y_j\}_{j\in\mathbb{Z}}\in X$ be a maximal collection of points satisfying

$$\frac{1}{2}\delta<\inf_{j\neq k}\rho(y_j,y_k)\le\delta. \tag{2.8}$$

By the maximality of $\{y_j\}_{j\in\mathbb{Z}}$, we have that for each $x\in X$ there exists a point y_j such that $\rho(x,y_j)\le\delta$. Let $\eta_j(y)=\theta(\frac{\rho(y,y_j)}{\delta})$ and $\bar\eta_j(y)=\big[\sum_i\eta_i(y)\big]^{-1}\eta_j(y)$. To see that $\bar\eta_j$ is well defined, it suffices to show that for any $y\in X$, there are only finitely many η_j with $\eta_j(y)\neq 0$. This follows from the following fact: $\eta_j(y)\neq 0$ if and only if $\rho(y,y_j)\le 2\delta$ and hence this implies that $B(y_j,\delta)\subseteq B(y,4A\delta)$. Inequalities (2.8) show $B(y_j,\frac{\delta}{4A})\cap B(y_k,\frac{\delta}{4A})=\phi$, and thus there are at most CA points $y_j\in X$ such that $B(y_j,\frac{\delta}{4A})\subseteq B(y,4A\delta)$. Now let $\Gamma=\{j:\bar\eta_j(y)\chi_0(y)\neq 0\}$. Note that $\#\Gamma\le Cr\delta$ since $\mu(\text{supp}\chi_0)\sim r$ and $\mu(\text{supp}\bar\eta_j)\sim\delta$. We write

$$\lambda_\delta(x,y)\chi_0(y)[\phi(y)-\phi(x)]\psi(x)=\sum_{j\in\Gamma}\lambda_\delta(x,y)\bar\eta_j(y)\chi_0(y)[\phi(y)-\phi(x)]\psi(x),$$

and we obtain

$$\begin{aligned}&\langle K(x,y),\lambda_\delta(x,y)\chi_0(y)[\phi(y)-\phi(x)]\psi(x)\rangle\\&\quad=\sum_{j\in\Gamma}\langle K(x,y),\lambda_\delta(x,y)\bar\eta_j(y)\chi_0(y)[\phi(y)-\phi(x)]\psi(x)\rangle.\end{aligned}$$

It is then easy to check that supp$\{\lambda_\delta(x,y)\bar\eta_j(y)\chi_0(y)[\phi(y)-\phi(x)]\psi(x)\}\subseteq B(y_j,3A\delta)\times B(y_j,2\delta)$ and

$$\|\lambda_\delta(x,y)\bar\eta_j(y)\chi_0(y)[\phi(y)-\phi(x)]\psi(x)\|_\infty\le C\delta^\eta$$

where C is a constant depending only on θ,ϕ,ψ,x_0, and r but not on δ and j.

We claim that

$$\|\lambda_\delta(.,y)\bar\eta_j(y)\chi_0(y)[\phi(y)-\phi(.)]\psi(.)\|_\eta\le C, \tag{2.9}$$

and

$$\|\lambda_\delta(x,.)\bar\eta_j(.)\chi_0(.)[\phi(.)-\phi(x)]\psi(x)\|_\eta\le C. \tag{2.10}$$

We accept (2.9) and (2.10) for the moment. Then, since T satisfies the "strong" weak boundedness property, we have

$$\begin{aligned}
&|\langle K(x,y), \lambda_\delta(x,y)\chi_0(y)[\phi(y)-\phi(x)]\psi(x)\rangle| \\
&\quad \le \sum_{j\in\Gamma} |\langle K(x,y), \lambda_\delta(x,y)\bar{\eta}_j(y)\chi_0(y)[\phi(y)-\phi(x)]\psi(x)\rangle| \\
&\quad \le \sum_{j\in\Gamma} C\mu(B(y_j, 3A\delta))\delta^\eta \le C\frac{r}{\delta}CA\delta\delta^\eta = CAr\delta^\eta
\end{aligned}$$

which yields (2.6).

It remains to show (2.9) and (2.10). We prove only (2.9) since the proof of (2.10) is similar. To show (2.9) it suffices to show that for $x, x_1 \in X$ and $\rho(x, x_1) \le \delta$,

$$\begin{aligned}
&|\bar{\eta}_j(y)\chi_0(y)||\lambda_\delta(x,y)[\phi(y)-\phi(x)]\psi(x) - \lambda_\delta(x_1,y)[\phi(y)-\phi(x_1)]\psi(x_1)| \\
&\quad \le C\rho(x,x_1)^\eta,
\end{aligned}$$

since if $\rho(x, x_1) \ge \delta$, then the expansion on the left above is clearly bounded by

$$\begin{aligned}
&|\bar{\eta}_j(y)\chi_0(y)|\{|\lambda_\delta(x,y)[\phi(y)-\phi(x)]\psi(x)| + |\lambda_\delta(x_1,y)[\phi(y)-\phi(x_1)]\psi(x_1)|\} \\
&\quad \le C\delta^\eta \le C\rho(x,x_1)^\eta.
\end{aligned}$$

By the construction of $\bar{\eta}_j$, it follows that

$$|\bar{\eta}_j(y)\chi_0(y)| \le C$$

for all $y \in X$. Thus

$$\begin{aligned}
&|\bar{\eta}_j(y)\chi_0(y)||\lambda_\delta(x,y)[\phi(y)-\phi(x)]\psi(x) - \lambda_\delta(x_1,y)[\phi(y)-\phi(x_1)]\psi(x_1)| \\
&\quad \le C|\lambda_\delta(x,y)[\phi(y)-\phi(x)]\psi(x) - \lambda_\delta(x_1,y)[\phi(y)-\phi(x_1)]\psi(x_1)| \\
&\quad \le C|[\lambda_\delta(x,y)-\lambda_\delta(x_1,y)][\phi(y)-\phi(x)]\psi(x)| \\
&\qquad +|\lambda_\delta(x_1,y)[\phi(x)-\phi(x_1)]\psi(x)| \\
&\qquad +|\lambda_\delta(x_1,y)[\phi(y)-\phi(x_1)][\psi(x)-\psi(x_1)]| \\
&\quad := I + II + III.
\end{aligned}$$

Recall that $\rho(x, x_1) \le \delta$. If $\rho(x, y) > C\delta$, where C is a constant depending on A but not on δ, then $\lambda_\delta(x, y) = \lambda_\delta(x_1, y) = 0$, so $I = 0$. Thus we may assume that $\rho(x, y) \le C\delta$ and with θ in (1.7),

$$\begin{aligned}
I &\le C\left|\frac{\rho(x,y)}{\delta} - \frac{\rho(x_1,y)}{\delta}\right|\rho(x,y)^\eta \le C\delta^{\eta-1}\rho(x,x_1)^\theta[\rho(x,y)+\rho(x_1,y)]^{1-\theta} \\
&\le C\delta^{\eta-\theta}\rho(x,x_1)^\theta \le C\rho(x,x_1)^\eta
\end{aligned}$$

since we may assume $\eta \leq \theta$. Terms II and III are easy to estimate:

$$II \leq C\rho(x, x_1)^{\eta},$$

$$III \leq C\rho(x, x_1)^{\eta},$$

since we can assume that $\delta < 1$. This completes the proof of (2.9) and implies

$$|p| \leq C\|\psi\|_1.$$

To finish the proof of Lemma 2.5, we now estimate q. It suffices to show that for $x \in B(x_0, r)$,

$$|T\chi_0(x)| \leq C. \tag{2.11}$$

To see this, it is easy to check that $q = \langle T\chi_0, \phi\psi\rangle$, and hence (2.10) implies

$$|q| \leq \|T\chi_0\|_{L^\infty(B(x_0,r))}\|\phi\psi\|_{L^1(B(x_0,r))} \leq C\|\psi\|_1.$$

To show (2.11), we use Meyer's idea again ([M1]). Let $\psi \in C^{\eta}(X)$ with supp$\psi \subseteq B(x_0, r)$ and $\int \psi(x)d\mu(x) = 0$. By the facts that $T(1) = 0, \int \psi(x)d\mu(x) = 0$, and the conditions on K, we obtain

$$\begin{aligned}|\langle T\chi_0, \psi\rangle| = |-\langle T\chi_1, \psi\rangle| &= \left|\iint [K(x,y) - K(x_0,y)]\chi_1(y)\psi(x)d\mu(x)d\mu(y)\right| \\ &\leq C\|\psi\|_1.\end{aligned}$$

Thus, $T\chi_0(x) = \omega + \gamma(x)$ for $x \in B(x_0, r)$, where ω is a constant and $\|\gamma\|_\infty \leq C$. To estimate ω, choose $\phi_1 \in C_0^{\eta}(X)$ with supp $\phi_1 \subseteq B(x_0, r)$, $\|\phi_1\|_\infty \leq 1$, $\|\phi_1\|_\eta \leq r^{-\eta}$ and $\int \phi_1(x)d\mu(x) = Cr$. We then have, by the "strong" weak boundedness property of T,

$$\left|Cr\omega + \int \phi_1(x)\gamma(x)d\mu(x)\right| = |\langle T\chi_0, \phi_1\rangle| \leq Cr$$

which implies $|\omega| \leq C$ and hence Lemma 2.5. ∎

We remark that the calculation above, together with the dominated convergence theorem and $T1 = 0$, yields the following integral representation:

$$\begin{aligned}&\langle T\phi, \psi\rangle \\ &\quad = \int_\Omega K(x,y)\{\chi_0(y)[\phi(y) - \phi(x)] - \chi_1(y)\phi(x)\}\psi(x)d\mu(y)d\mu(x)\end{aligned} \tag{2.12}$$

and

$$\begin{aligned}&\langle K(x,y), [\phi(y) - \phi(x)]\chi_0(y)\rangle \\ &\quad = \lim_{\delta\to 0}\int_{\rho(x,y)\geq\delta} K(x,y)\chi_0(y)[\phi(y) - \phi(x)]d\mu(y)\end{aligned} \tag{2.13}$$

where χ_0, ϕ and ψ are defined as above.

We return to prove the Theorem 2.4. Fix a function $\theta \in C^1(\mathbb{R})$ with supp $\theta \subseteq \{x \in \mathbb{R} : |x| \leq 2\}$ and $\theta = 1$ on $\{x \in \mathbb{R} : |x| \leq 1\}$. Suppose that $f \in \mathcal{M}_0(x_0, r, \beta, \gamma)$ with $x_0 \in X, r > 0$ and $0 < \beta, \gamma < \epsilon$. We first prove that $T(f)(x)$ satisfies the size condition (i) of Definition 2.1. To do this, we first consider the case where $\rho(x, x_0) \leq 5r$. Set $1 = \xi(y) + \eta(y)$ where $\xi(y) = \theta(\frac{\rho(y,x_0)}{10Ar})$. Then we have

$$T(f)(x) = \int K(x,y)\xi(y)[f(y) - f(x)]d\mu(y) + \int K(x,y)\eta(y)f(y)d\mu(y)$$
$$+f(x)\int K(x,y)\xi(y)d\mu(y) := I + II + III.$$

Using (2.13),

$$\begin{aligned}|I| &\leq C\int_{\rho(x,y)\leq 25A^2r} |K(x,y)||f(y) - f(x)|d\mu(y)\\ &\leq C\|f\|_{\mathcal{M}(x_0,r,\beta,\gamma)}\int_{\rho(x,y)\leq 25A^2r} \rho(x,y)^{-1}\frac{\rho(x,y)^\beta}{r^{1+\beta}}d\mu(y)\\ &\leq C\|f\|_{\mathcal{M}(x_0,r,\beta,\gamma)}r^{-1}.\end{aligned}$$

By Lemma 2.5,

$$|III| \leq C|f(x)| \leq C\|f\|_{\mathcal{M}(x_0,r,\beta,\gamma)}r^{-1}.$$

For term II we have

$$\begin{aligned}|II| &\leq C\|f\|_{\mathcal{M}(x_0,r,\beta,\gamma)}\int_{\rho(x,y)\geq 10Ar} \rho(x,y)^{-1}\frac{r^\gamma}{\rho(y,x_0)^{1+\gamma}}d\mu(y)\\ &\leq C\|f\|_{\mathcal{M}(x_0,r,\beta,\gamma)}r^{-1}\end{aligned}$$

since $\rho(x, x_0) \leq 5r$.

This implies that $T(f)(x)$ satisfies (i) of Definition 2.1 with $\rho(x, x_0) \leq 5r$. Consider now $\rho(x, x_0) = R > 5r$. Following the proof in [M1], set $1 = I(y) + J(y) + L(y)$, where $I(y) = \theta(\frac{4A\rho(y,x)}{R}), J(y) = \theta(\frac{4A\rho(y,x_0)}{R})$, and $f_1(y) = f(y)I(y), f_2(y) = f(y)J(y)$, and $f_3(y) = f(y)L(y)$. Then it is easy to check the following estimates:

$$|f_1(y)| \leq C\|f\|_{\mathcal{M}(x_0,r,\beta,\gamma)}\frac{r^\gamma}{R^{1+\gamma}}; \tag{2.14}$$

$$\begin{aligned}|f_1(y) - f_1(y')| &\leq |I(y)||f(y) - f(y')| + |f(y')||I(y) - I(y')| \\ &\leq C\|f\|_{\mathcal{M}(x_0,r,\beta,\gamma)}\frac{\rho(y,y')^\beta}{R^\beta}\frac{r^\gamma}{R^{1+\gamma}}\end{aligned} \tag{2.15}$$

for all y and y';

$$|f_3(y)| \leq C\|f\|_{\mathcal{M}(x_0,r,\beta,\gamma)} \frac{r^\gamma}{\rho(y,x_0)^{1+\gamma}} \chi_{\{y\in X:\rho(y,x_0)>\frac{1}{4A}R\}}; \tag{2.16}$$

$$\int |f_3(y)|d\mu(y) \leq C\|f\|_{\mathcal{M}(x_0,r,\beta,\gamma)} \frac{r^\gamma}{R^\gamma}; \tag{2.17}$$

$$\begin{aligned} \left|\int f_2(y)d\mu(y)\right| &= \left| -\int f_1(y)d\mu(y) - \int f_3(y)d\mu(y)\right| \\ &\leq C\|f\|_{\mathcal{M}(x_0,r,\beta,\gamma)} \frac{r^\gamma}{R^\gamma}. \end{aligned} \tag{2.18}$$

We write

$$\begin{aligned} T(f_1)(x) &= \int K(x,y)u(y)[f_1(y)-f_1(x)]d\mu(y) + f_1(x)\int K(x,y)u(y)d\mu(y) \\ &= \sigma_1(x) + \sigma_2(x) \end{aligned}$$

where $u(y) = \theta(\frac{2A\rho(x,y)}{R})$. Applying the estimate (2.15) and Lemma 2.5, we obtain

$$\begin{aligned} |\sigma_1(x)| &\leq C\|f\|_{\mathcal{M}(x_0,r,\beta,\gamma)} \int_{\rho(x,y)\leq\frac{R}{A}} \rho(x,y)^{-1} \frac{\rho(x,y)^\beta}{R^\beta} \frac{r^\gamma}{R^{1+\gamma}} d\mu(y) \\ &\leq C\|f\|_{\mathcal{M}(x_0,r,\beta,\gamma)} \frac{r^\gamma}{R^{1+\gamma}}; \end{aligned}$$

and

$$|\sigma_2(x)| \leq C|f_1(x)| \leq C\|f\|_{\mathcal{M}(x_0,r,\beta,\gamma)} \frac{r^\gamma}{R^{1+\gamma}}.$$

Notice that x is not in the support of f_2. We can write

$$\begin{aligned} T(f_2)(x) &= \int [K(x,y)-K(x,x_0)]f_2(y)d\mu(y) + K(x,x_0)\int f_2(y)d\mu(y) \\ &= \delta_1(x) + \delta_2(x). \end{aligned}$$

Using the estimates on the kernel of T and on f_2 in (2.18), we then get

$$\begin{aligned} |\delta_1(x)| &\leq C\|f\|_{\mathcal{M}(x_0,r,\beta,\gamma)} \int_{\rho(x_0,y)\leq\frac{R}{2A}} \frac{\rho(x_0,y)^\epsilon}{R^{1+\epsilon}} \frac{r^\gamma}{\rho(x_0,y)^{1+\gamma}} d\mu(y) \\ &\leq C\|f\|_{\mathcal{M}(x_0,r,\beta,\gamma)} \frac{r^\gamma}{R^{1+\gamma}} \end{aligned}$$

since $\gamma < \epsilon$, and

$$|\delta_2(x)| \leq CR^{-1}\left|\int f_2(y)d\mu(y)\right| \leq C\|f\|_{\mathcal{M}(x_0,r,\beta,\gamma)} \frac{r^\gamma}{R^{1+\gamma}}.$$

Finally, since x is not in the support of f_3, (2.16) implies

$$|T(f_3)(x)| \leq C\|f\|_{\mathcal{M}(x_0,r,\beta,\gamma)} \int_{\rho(x,y)\geq \frac{R}{4A}, \rho(x_0,y)\geq \frac{R}{4A}} \rho(x,y)^{-1} \frac{r^\gamma}{\rho(x_0,y)^{1+\gamma}} d\mu(y)$$
$$\leq C\|f\|_{\mathcal{M}(x_0,r,\beta,\gamma)} \frac{r^\gamma}{R^{1+\gamma}}.$$

This yields that $T(f)(x)$ satisfies (i) of Definition 2.1 for $\rho(x,x_0) > 5r$ and hence, estimate (i) of Definition 2.1 for all $x \in X$.

Now we prove that $T(f)(x)$ satisfies the smoothness condition (ii) of Definition 2.1. To do this, set $\rho(x,x_0) = R$ and $\rho(x,x') = \delta$. We consider first the case where $R \geq 10r$ and $\delta \leq \frac{1}{20A^2}(r+R)$. As in the above, set $1 = I(y) + J(y) + L(y)$, where $I(y) = \theta(\frac{8A\rho(y,x)}{R})$, $J(y) = \theta(\frac{8A\rho(y,x_0)}{R})$, and $f_1(y) = f(y)I(y)$, $f_2(y) = f(y)J(y)$, and $f_3(y) = f(y)L(y)$. We write

$$T(f_1)(x) = \int K(x,y)u(y)[f_1(y) - f_1(x)]d\mu(y)$$
$$+ \int K(x,y)v(y)f_1(y)d\mu(y) + f_1(x) \int K(x,y)u(y)d\mu(y)$$

where $u(y) = \theta(\frac{\rho(x,y)}{2A\delta})$ and $v(y) = 1 - u(y)$. Denote the first term of the above right-hand side by $p(x)$ and the last two terms by $q(x)$. The size condition of K and the smoothness of f_1 in (2.15) yield

$$|p(x)| \leq C\|f\|_{\mathcal{M}(x_0,r,\beta,\gamma)} \int_{\rho(x,y)\leq 4A\delta} \rho(x,y)^{-1} \frac{\rho(x,y)^\beta}{R^\beta} \frac{r^\gamma}{R^{1+\gamma}} d\mu(y)$$
$$\leq C\|f\|_{\mathcal{M}(x_0,r,\beta,\gamma)} \frac{\delta^\beta}{R^\beta} \frac{r^\gamma}{R^{1+\gamma}}.$$

This estimate still holds with x replaced by x' for $\rho(x,x') = \delta$. Thus

$$|p(x) - p(x')| \leq C\|f\|_{\mathcal{M}(x_0,r,\beta,\gamma)} \frac{\delta^\beta}{R^\beta} \frac{r^\gamma}{R^{1+\gamma}}.$$

For $q(x)$, using the condition $T1 = 0$, we obtain

$$q(x) - q(x') = \int [K(x,y) - K(x',y)]v(y)[f_1(y) - f_1(x)]d\mu(y)$$
$$+[f_1(y) - f_1(x)] \int K(x,y)u(y)d\mu(y)$$
$$= I + II.$$

Using Lemma 2.5 and the estimate for f_1 in (2.15),

$$|II| \leq C|f_1(x) - f_1(x')| \leq C\|f\|_{\mathcal{M}(x_0,r,\beta,\gamma)} \frac{\delta^\beta}{R^\beta} \frac{r^\gamma}{R^{1+\gamma}}.$$

Observing

$$|f_1(y) - f_1(x)||v(y)| \le C\|f\|_{\mathcal{M}(x_0,r,\beta,\gamma)} \frac{\rho(x,y)^\beta}{R^\beta} \frac{r^\gamma}{R^{1+\gamma}}$$

for all $y \in X$, we see that I is dominated by

$$C \int_{\rho(x,y)\ge 2A\delta} |K(x,y) - K(x',y)||v(y)||f_1(y) - f_1(x)|d\mu(y)$$

$$\le C\|f\|_{\mathcal{M}(x_0,r,\beta,\gamma)} \int_{\rho(x,y)\ge 2A\delta} \frac{\rho(x,x')^\epsilon}{\rho(x,y)^{1+\epsilon}} \frac{\rho(x,y)^\beta}{R^\beta} \frac{r^\gamma}{R^{1+\gamma}} d\mu(y)$$

$$\le C\|f\|_{\mathcal{M}(x_0,r,\beta,\gamma)} \frac{\delta^\beta}{R^\beta} \frac{r^\gamma}{R^{1+\gamma}}$$

since $\beta < \epsilon$. This implies

$$|T(f_1)(x) - T(f_1)(x')| \le C\|f\|_{\mathcal{M}(x_0,r,\beta,\gamma)} \frac{\delta^\beta}{R^\beta} \frac{r^\gamma}{R^{1+\gamma}}.$$

Note that for $\rho(x,x') = \delta \le \frac{1}{20A^2}(r+R)$ and $R \ge 10r, x$ and x' are not in the supports of f_2 and f_3. Using the condition for K and the estimate for f_2 in (2.18), then

$$|T(f_2)(x) - T(f_2)(x')| = \left| \int [K(x,y) - K(x',y)] f_2(y) d\mu(y) \right|$$

$$\le \int |K(x,y) - K(x',y) - K(x,x_0) - K(x',x_0)||f_2(y)|d\mu(y)$$

$$+|K(x,x_0) - K(x',x_0)| \left| \int f_2(y) d\mu(y) \right|$$

$$\le C\|f\|_{\mathcal{M}(x_0,r,\beta,\gamma)} \left\{ \int_{\rho(x_0,y)\le \frac{R}{4A}} \frac{\rho(x,x')^\epsilon \rho(y,x_0)^\epsilon}{R^{2+\epsilon}} \frac{r^\gamma}{\rho(y,x_0)^{1+\gamma}} d\mu(y) + \frac{\delta^\epsilon}{R^{1+\epsilon}} \frac{r^\gamma}{R^\gamma} \right\}$$

$$\le C\|f\|_{\mathcal{M}(x_0,r,\beta,\gamma)} \frac{\delta^\epsilon}{R^\epsilon} \frac{r^\gamma}{R^{1+\gamma}}$$

since $\gamma < \epsilon$. Finally, we have

$$|T(f_3)(x) - T(f_3)(x')| = \left| \int [K(x,y) - K(x',y)] f_3(y) d\mu(y) \right|$$

$$\le C \int_{\rho(x,y)\ge \frac{R}{8A} \ge 2A\delta} \frac{\rho(x,x')^\epsilon}{\rho(x,y)^{1+\epsilon}} |f_3(y)|d\mu(y) \le C\|f\|_{\mathcal{M}(x_0,r,\beta,\gamma)} \frac{\delta^\epsilon}{R^\epsilon} \frac{r^\gamma}{R^{1+\gamma}}.$$

These estimates imply that $T(f)(x)$ satisfies the condition (ii) of Definition 2.1 for the case where $\rho(x,x_0) = R \ge 10r$ and $\rho(x,x') = \delta \le \frac{1}{20A^2}(r+R)$. We now consider the other cases. Note first that if $\rho(x,x_0) = R$ and $\frac{1}{2A}(r+R) \ge$

$\rho(x, x') = \delta \geq \frac{1}{20A^2}(r+R)$, then the estimate (ii) of Definition 2.1 for $T(f)(x)$ follows from the estimate (i) of Definition 2.1 for $T(f)(x)$. So we only need to consider the case where $R \leq 10r$ and $\delta \leq \frac{1}{20A^2}(r+R)$. This case is similar and easier. In fact, all we need to do is to replace R in the proof above by r. We leave these details to the reader. The proof of Theorem 2.4 is completed.

We remark that the condition in (2.2) is also necessary for the boundedness of Calderón-Zygmund operators on *wavelet spaces*. To be precise, in the next chapter, we will prove all kinds of Calderón's identities and use them to provide all kinds of wavelet expansions of functions and distributions on spaces of homogeneous type. Suppose that T is a Calderón-Zygmund operator and maps the wavelet space $\mathcal{M}_0(x_0, r, \beta, \gamma)$ to itself. By the wavelet expansion given in Theorem 3.25 below, $K(x, y)$, the kernel of T, can be written as $K(x, y) = \sum_{\lambda \in \Lambda} T(\widetilde{\psi}_\lambda)(x)\psi_\lambda(y)$. Since $\widetilde{\psi}_\lambda(x)$ is a wavelet, by the assumption on $T, T(\widetilde{\psi}_\lambda)(x)$ is also a wavelet. Then one can easily check that $K(x, y)$ satisfies the condition (2.2) but the exponent ϵ must be replaced by ϵ' with $0 < \epsilon' < \beta, \gamma$. We leave these details to the reader.

Chapter 3
Wavelet Expansions on Spaces of Homogeneous Type

3.1 Introduction

Up to now, we succeeded in building a Littlewood-Paley analysis on a space of homogeneous type developed by G. David, J. L. Journé and S. Semmes, and in proving the boundedness of a class of Calderón-Zygmund operators on wavelet spaces. Now we aim at bridging the gap between operator theory and wavelets. We will use the theory of Calderón-Zygmund operators to derive a wavelet expansion from a Littlewood-Paley analysis. For doing it we recall the Littlewood-Paley analysis developed by G. David, J. L. Journé and S. Semmes, which has been described in Chapter 1. Let S_k and D_k be operators defined by Coifman's construction. G. David, J. L. Journé and S. Semmes provided the following identity: For $f \in L^2$,

$$f = \sum_k (T_N)^{-1} D_k^N D_k(f)$$

where the operator T_N is invertible on L^2 for large N and $(T_N)^{-1}$ is bounded on L^2 uniformly for large N.

As we explained before, this identity is different from Calderón's identity. But, however, it is a starting point leading to Calderón's identity on spaces of homogeneous type. In Section 3.3, we introduce the definition of an approximation to the identity and give basic estimates on the kernels of an approximation to the identity. One can easily check that $D_k^N(x,y)$, the kernels of D_k^N, are wavelets in the sense given in the Introduction. In Section 3.4, we will prove the fundamental fact that $(T_N)^{-1}$ is bounded on wavelet spaces. Let $\widetilde{D}_k = (T_N)^{-1} D_k^N$. Then Calderón's identity on spaces of homogeneous type is given by:

$$f = \sum_k \widetilde{D}_k D_k(f)$$

where the kernels of $\widetilde{D}_k$ are wavelets in the sens given in the Introduction.

D. Deng and Y. Han, *Harmonic Analysis on Spaces of Homogeneous Type*, Lecture Notes in Mathematics 1966,

By a discrete version of the above identity, we obtain one of main results in this book, that is, Theorem 3.25 below. To highlight the relation between Theorem 3.25 and wavelet expansions on spaces of homogeneous type, let us return to the classical setting. If $\psi_{j,k}(x) = 2^{\frac{nj}{2}}(2^j x - k), j \in \mathbb{Z}, k \in \mathbb{Z}^n, \psi \in F \subset \mathscr{S}(R^n)$ is an orthonormal wavelet basis, we have

$$f(x) = \sum_j \sum_k \langle f, \psi_{j,k} \rangle \psi_{j,k}(x) \tag{3.1}$$

(the sum over the finite set F consisting of the $2^n - 1$ mother wavelets ψ is omitted).

Let $\Delta_j(f)$ be $f * \psi_j$ where $\psi_j(x) = 2^{nj}\psi(2^j x)$ and let $D_j(x,y) = 2^{nj}\psi(2^j(x-y))$ be the kernel of Δ_j. Then we can write

$$\psi_{j,k}(x) = 2^{-\frac{nj}{2}} D_j(x, k2^{-j}) \tag{3.2}$$

which is ridiculous but paves the way to Theorem 3.25. We also have

$$2^{\frac{nj}{2}} \langle f, \psi_{j,k} \rangle = (\Delta_j^* f)(k2^{-j}) \tag{3.3}$$

where Δ_j^* is the adjoint of Δ_j.

Finally (3.1) can be rewritten as

$$f(x) = \sum_j \sum_k 2^{-nj} D_j(x, k2^{-j})(\Delta_j^* f)(k2^{-j}) \tag{3.4}$$

which is exactly Theorem 3.25.

But conversely (and this is far more important) Theorem 3.25 yields a wavelet series expansion. This is the true meaning of Theorem 3.25. The same remark applies to Theorem 3.27. Theorem 3.25 says that

$$f = \sum_{\lambda \in \Lambda} \langle f, \psi_\lambda \rangle \widetilde{\psi}_\lambda \tag{3.5}$$

while Theorem 3.27 is a variant where one does not use "large wavelets" since these wavelets sum up to the scaling function. The proofs of Theorem 3.25 and Theorem 3.27 will be given in Section 3.5.

Orthonormal wavelets bases are out of reach on a space of homogeneous type. Instead the theory of frames will be used. Roughly speaking using a frame means that you tolerate a limited amount of redundancy while redundancy is compeletely avoided with a basis. For reader's convenience, some few facts about frames will be proved now in the next section.

3.2 The Theory of Frames

Let H be a Hilbert space and let $e_j, j \in J$, be a collection of vectors in H. This collection is a frame if and only if there exists a constant $C \geq 1$ such that, for every x in H, we have

$$\frac{1}{C}\|x\|^2 \leq \sum_{j\in J} |\langle x, e_j\rangle|^2 \leq C\|x\|^2. \tag{3.6}$$

This is equivalent to saying that the self-adjoint operator

$$S(x) = \sum_{j\in J} \langle x, e_j\rangle e_j \tag{3.7}$$

has the following property

$$\frac{1}{C}\mathbb{I} \leq S \leq C\mathbb{I}. \tag{3.8}$$

where $\mathbb{I}$ is the identity operator. Therefore S is invertible and x can be recovered through

$$x = \sum_{j\in J} \langle x, e_j\rangle f_j \tag{3.9}$$

where $f_j = S^{-1}(e_j)$.

Theorem 3.1. *The following two properties are equivalent ones:*
(a) e_j, $j \in J$, is a frame in H.
(b) The operator $\Lambda : \ell^2(J) \to H$ defined by $\Lambda(\alpha_j) = \sum_j \alpha_j e_j$ is continuous and onto.

We first prove that (a) implies (b). The continuity of Λ is proved by the following way:

$$\begin{aligned}\left\|\sum \alpha_j e_j\right\| &= \sup_{\|x\|\leq 1} \left|\left\langle \sum \alpha_j e_j, x\right\rangle\right| \\ &= \sup_{\|x\|\leq 1} \left|\sum \alpha_j \overline{\langle x, e_j\rangle}\right| \\ &\leq C\left(\sum \left|\alpha_j\right|^2\right)^{\frac{1}{2}}\end{aligned}$$

since $\sum |\langle x, e_j\rangle|^2 \leq C$ by (3.6).

We already proved that Λ is onto. We now turn to the converse implication.

If Λ is onto, Banach's theorem says that there exists a constant C such that for each $x \in H$ one can find $\alpha_j \in \ell^2(J)$ with $x = \sum \alpha_j e_j$ and $\sum |\alpha_j|^2 \le C^2\|x\|^2$. Therefore

$$\begin{aligned}\|x\|^2 = \langle x, x\rangle = \Big\langle \sum \alpha_j e_j, x\Big\rangle = \sum \alpha_j \overline{\langle x, e_j\rangle} \\ \le \Big(\sum |\alpha_j|^2\Big)^{\frac{1}{2}} \Big(\sum |\langle x, e_j\rangle|^2\Big)^{\frac{1}{2}} \\ \le C\|x\| \Big(\sum |\langle x, e_j\rangle|^2\Big)^{\frac{1}{2}}.\end{aligned}$$

It implies $\|x\| \le C(\sum |\langle x, e_j\rangle|^2)^{\frac{1}{2}}$. This is the "difficult half" of (3.6). The other implication is simpler. We have

$$\Big(\sum_{j\in J} |\langle x, e_j\rangle|^2\Big)^{\frac{1}{2}} = \sup_{\|\alpha_j\|_2 \le 1} \Big|\sum \alpha_j \langle x, e_j\rangle\Big|.$$

But $\sum \alpha_j \langle x, e_j\rangle = \langle x, \sum \overline{\alpha_j}, e_j\rangle$ and the continuity of Λ implies the required conclusion.

Definition 3.2. We say that $(e_j)_{j\in J}$ is almost orthogonal in H if a constant C exists such that

$$\Big\|\sum_{j\in J} \alpha_j e_j\Big\| \le C\Big(\sum_{j\in J} |\alpha_j|^2\Big)^{\frac{1}{2}} \tag{3.10}$$

for every sequence $(\alpha_j)_{j\in J} \in \ell^2(J)$.

This is equivalent to

$$\sum_{j\in J} |\langle x, e_j\rangle|^2 \le \|x\|^2 \tag{3.11}$$

for all $x \in H$.

Theorem 3.3. *Let $(e_j)_{j\in J}$ and $(f_j)_{j\in J}$ be two families of vectors in H. Let us assume that both $(e_j)_{j\in J}$ and $(f_j)_{j\in J}$ satisfy (3.10). Then the operator $S : H \to H$ defined by*

$$S(x) = \sum_{j\in J} \langle x, e_j\rangle f_j \tag{3.12}$$

is continuous from H to H. If S is one to one, then both $(e_j)_{j\in J}$ and $(f_j)_{j\in J}$ are frames.

The proof is fairly obvious. We have for a positive constant γ,

$$\gamma\|x\| \leq \|S(x)\| \leq C\Big(\sum |\langle x, e_j\rangle|^2\Big)^{\frac{1}{2}} \tag{3.13}$$

by (3.10) applied to the family $(f_j)_{j\in J}$. Therefore $(e_j)_{j\in J}$ is a frame. But the same argument applies to S^* and $(f_j)_{j\in J}$ is also a frame.

One should observe that the frame expansion given in (3.9) is not a wavelet expansion because, in general, f_j are not wavelets even e_j could be wavelets. For our purpose, however, we will provide a wavelet expansion. This leads to add some extra conditions on frames. We describe these conditions now in the next section.

3.3 Approximation to the Identity and Basic Estimates

We now define an approximation to the identity on spaces of homogeneous type. This generalizes the properties of S_k in Section 1.3.

Definition 3.4. A sequence $\{S_k\}_{k\in\mathbb{Z}}$ of operators is said to be an *approximation to the identity* if there exist $0 < \sigma, \epsilon \leq \theta$ and $C < \infty$ such that for all x, x', y and $y' \in X$, $S_k(x,y)$, the kernel of S_k, are functions from $X \times X$ into $\mathbb{C}$ satisfying

$$|S_k(x,y)| \leq C\frac{2^{-k\epsilon}}{(2^{-k}+\rho(x,y))^{1+\epsilon}}; \tag{3.14}$$

$$|S_k(x,y) - S_k(x',y)| \leq C\Big(\frac{\rho(x,x')}{2^{-k}+\rho(x,y)}\Big)^{\sigma}\frac{2^{-k\epsilon}}{(2^{-k}+\rho(x,y))^{1+\epsilon}} \tag{3.15}$$

for $\rho(x,x') \leq \frac{1}{2A}(2^{-k}+\rho(x,y)$;

$$|S_k(x,y) - S_k(x,y')| \leq C\Big(\frac{\rho(y,y')}{2^{-k}+\rho(x,y)}\Big)^{\sigma}\frac{2^{-k\epsilon}}{(2^{-k}+\rho(x,y))^{1+\epsilon}} \tag{3.16}$$

for $\rho(y,y') \leq \frac{1}{2A}(2^{-k}+\rho(x,y)$;

$$\int S_k(x,y)d\mu(y) = 1 \tag{3.17}$$

for all $x \in X$;

$$\int S_k(x,y)d\mu(x) = 1 \tag{3.18}$$

for all $y \in X$.

One should observe that if S_k is an approximation to the identity, $S_k(x,y)$, the kernel of S_k, as a function of x when y is fixed, or a function of y when x is fixed, is a scaling function (see Definition 2.1 and the remarks following this definition). For our purpose, we also need the following double Lipschitz condition for an approximation to the identity . This condition is crucial for providing wavelet expansions on spaces of homogeneous type.

Definition 3.5. An approximation to the identity $\{S_k\}_{k\in\mathbb{Z}}$ is said to satisfy the double Lipschitz condition if

$$\begin{aligned}&|S_k(x,y)-S_k(x',y)-S_k(x,y')+S_k(x',y')| \\ &\leq C\Big(\frac{\rho(x,x')}{2^{-k}+\rho(x,y)}\Big)^{\sigma}\Big(\frac{\rho(y,y')}{2^{-k}+\rho(x,y)}\Big)^{\sigma}\frac{2^{-k\epsilon}}{(2^{-k}+\rho(x,y))^{1+\epsilon}}\end{aligned} \tag{3.19}$$

for $\rho(x,x')\leq\frac{1}{2A}(2^{-k}+\rho(x,y))$ and $\rho(y,y')\leq\frac{1}{2A}(2^{-k}+\rho(x,y))$.

The following lemmas provide the basic estimates of approximations to the identity, which will be often used later.

Lemma 3.6. *Suppose that $S_k(x,y), k\in\mathbb{Z}$, the kernels of operators $S_k, k\in\mathbb{Z}$, satisfy the conditions (3.14), (3.15) and (3.17). Set $D_k=S_k-S_{k-1}$ for all $k\in\mathbb{Z}$. Then for any $\sigma'\leq\sigma$ and $\sigma'<\epsilon$, there exists a constant C which depends only on σ,σ' and ϵ but not on k and l, such that*

$$|D_kD_l(x,y)|\leq C(2^{-(k-l)\sigma'}\wedge 1)\frac{2^{-(k\wedge l)\epsilon}}{(2^{-(k\wedge l)}+\rho(x,y))^{1+\epsilon}} \tag{3.20}$$

where $D_kD_l(x,y)$ is the kernel of operator D_kD_l and $a\wedge b=\min\{a,b\}$.

To prove the estimate in (3.20), we write $D_kD_l(x,y)=\int D_k(x,z)D_l(z,y)$ $d\mu(z)$ and consider that $l\geq k$. In this case, we use only the size condition (3.14). Considering $\rho(x,y)\leq 4A2^{-k}$ and $\rho(x,y)>4A2^{-k}$, respectively, the estimate in (3.20) follows easily.

We now consider that $l\leq k$. In this case, $D_l(x,z)$, the kernel of D_l, as the function of z, is flat, and $D_k(z,y)$, the kernel of D_k, as the function of z, has more oscillations, so we use the cancellation condition (3.17) on D_k and the smoothness condition (3.15) on $D_l(z,y)$. We have

$$\begin{aligned}|D_kD_l(x,y)|&=\Big|\int D_k(x,z)[D_l(z,y)-D_l(x,y)]d\mu(z)\Big|\\ &\leq\int_{W_1}|D_k(x,z)||D_l(z,y)-D_l(x,y)|d\mu(z)\\ &\quad+\int_{W_2}|D_k(x,z)||D_l(z,y)-D_l(x,y)|d\mu(z)=I+II\end{aligned}$$

where $W_1=\{z\in X:\rho(z,x)\leq\frac{1}{2A}(2^{-l}+\rho(x,y))\}$ and $W_2=\{z\in X:\rho(z,x)\geq\frac{1}{2A}(2^{-l}+\rho(x,y))\}$.

Using the smoothness estimate (3.15) and the fact $\sigma' < \epsilon$,

$$I \leq C \int \frac{2^{-k\epsilon}}{(2^{-k} + \rho(x,z))^{1+\epsilon}} \Big(\frac{\rho(x,z)}{2^{-l} + \rho(x,y)} \Big)^{\sigma'} \frac{2^{-l\epsilon}}{(2^{-l} + \rho(x,y))^{1+\epsilon}} d\mu(z)$$
$$\leq C 2^{-(k-l)\sigma'} \frac{2^{-l\epsilon}}{(2^{-l} + \rho(x,y))^{1+\epsilon}}.$$

Applying the size estimate (3.14) to term II implies

$$II \leq \int_{W_2} |D_k(x,z)||D_l(z,y)| d\mu(z)$$
$$+ \int_{W_2} |D_k(x,z)||D_l(x,y)| d\mu(z)$$
$$\leq C \frac{2^{-k\epsilon}}{(2^{-l} + \rho(x,y))^{1+\epsilon}} \int_{W_2} \frac{2^{-l\epsilon}}{(2^{-l} + \rho(z,y))^{1+\epsilon}} d\mu(z)$$
$$+ C \frac{2^{-l\epsilon}}{(2^{-l} + \rho(x,y))^{1+\epsilon}} \int_{W_2} \frac{2^{-k\epsilon}}{\rho(x,z)^{1+\epsilon}} d\mu(z)$$
$$\leq C 2^{-(k-l)\epsilon} \frac{2^{-l\epsilon}}{(2^{-l} + \rho(x,y))^{1+\epsilon}}.$$

This yields (3.20).

As a consequence of Lemma 3.6, we obtain the following almost orthogonal estimate.

Lemma 3.7. *Suppose that $\{S_k\}_{k\in\mathbb{Z}}$ is an approximation to the identity . Set $D_k = S_k - S_{k-1}$ for all $k \in \mathbb{Z}$. Then for any $\sigma' \leq \sigma$ and $\sigma' < \epsilon$, there exists a constant C which depends only on σ, σ' and ϵ but not on k and l, such that*

$$|D_k D_l(x,y)| \leq C 2^{-|l-k|\sigma'} \frac{2^{-(k\wedge l)\epsilon}}{(2^{-(k\wedge l)} + \rho(x,y))^{1+\epsilon}}. \tag{3.21}$$

Remark 3.8. The estimate in (3.21) means that the kernels of an approximation to the identity satisfy the so-called almost orthogonal estimate. Suppose that $S_k(x,y), k \in \mathbb{Z}$, the kernels of operators S_k, satisfy the conditions (3.14), (3.16) and (3.17), and $P_k(x,y), k \in \mathbb{Z}$, the kernels of operators P_k, satisfy the conditions (3.14), (3.15) and (3.18). Set $D_k = S_k - S_{k-1}$ for all $k \in \mathbb{Z}$ and $E_k = P_k - P_{k-1}$ for all $k \in \mathbb{Z}$. Then for any $\sigma' \leq \sigma$ and $\sigma' < \epsilon$, there exists a constant C which depends only on σ, σ' and ϵ but not on k, such that

$$|D_k E_l(x,y)| \leq C 2^{-|l-k|\sigma'} \frac{2^{-(k\wedge l)\epsilon}}{(2^{-(k\wedge l)} + \rho(x,y))^{1+\epsilon}}.$$

In fact, to see this, note that in the proof of Lemma 3.7 only the moment and smoothness conditions of $D_k(x,y)$ on the variable y and of $E_l(x,y)$ on the variable x, are needed. This remark will be used later.

We now give the smoothness estimate on the kernel of D_kD_l.

Lemma 3.9. *Suppose that $D_k(x,y), k \in \mathbb{Z}$, the kernels of operators $D_k, k \in \mathbb{Z}$, satisfy the same conditions as in Lemma 3.6. Then there exists a constant C which depends only on σ, and ϵ but not on k, such that if $k \geq l$,*

$$|D_kD_l(x,y) - D_kD_l(x',y)| \leq C\Big(\frac{\rho(x,x')}{2^{-l}+\rho(x,y)}\Big)^{\sigma}\frac{2^{-l\epsilon}}{(2^{-l}+\rho(x,y))^{1+\epsilon}} \tag{3.22}$$

for $\rho(x,x') \leq \frac{1}{2A}(2^{-l}+\rho(x,y))$.

Proof. It suffices to consider only the case where $\rho(x,x') \leq \frac{1}{4A^2}(2^{-l}+\rho(x,y))$ because otherwise the estimate in (3.22) follows directly from Lemma 3.6. By the condition (3.17), $D_k(1) = 0$, and we get

$$\begin{aligned}
&|D_kD_l(x,y) - D_kD_l(x',y)| \\
&= \left|\int [D_k(x,z) - D_k(x',z)][D_l(z,y) - D_l(x,y)]d\mu(z)\right| \\
&\leq \int |D_k(x,z) - D_k(x',z)||D_l(z,y) - D_l(x,y)|d\mu(z) \\
&= \int_{W_1} \cdots + \int_{W_2} \cdots + \int_{W_3} \cdots = I + II + III
\end{aligned}$$

where

$$\begin{aligned}
W_1 &= \Big\{z \in X : \rho(x,x') \leq \frac{1}{4A^2}(2^{-l}+\rho(x,y)) \leq \frac{1}{2A}(2^{-k}+\rho(x,z))\Big\}, \\
W_2 &= \Big\{z \in X : \rho(x,x') \leq \frac{1}{2A}(2^{-k}+\rho(x,z)) \leq \frac{1}{4A^2}(2^{-l}+\rho(x,y))\Big\}, \\
W_3 &= \Big\{z \in X : \rho(x,x') > \frac{1}{2A}(2^{-k}+\rho(x,z))\Big\}.
\end{aligned}$$

Note that if $z \in W_2$, then $\rho(x,z)) \leq \frac{1}{2A}(2^{-l}+\rho(x,y))$. Using the estimate (3.15),

$$\begin{aligned}
II \leq C\int_{W_2} \Big(\frac{\rho(x,x')}{2^{-k}+\rho(x,z)}\Big)^{\sigma}\frac{2^{-k\epsilon}}{(2^{-k}+\rho(x,z))^{1+\epsilon}}\Big(\frac{\rho(z,x)}{2^{-l}+\rho(x,y)}\Big)^{\sigma} \\
\frac{2^{-l\epsilon}}{(2^{-l}+\rho(x,y))^{1+\epsilon}}d\mu(z) \leq C\Big(\frac{\rho(x,x')}{2^{-l}+\rho(x,y)}\Big)^{\sigma}\frac{2^{-l\epsilon}}{(2^{-l}+\rho(x,y))^{1+\epsilon}}.
\end{aligned}$$

The estimates (3.14) and (3.15) yield

$$
\begin{aligned}
I &\le C \int_{W_1} \Big(\frac{\rho(x,x')}{2^{-k}+\rho(x,z)}\Big)^{\sigma} \frac{2^{-k\epsilon}}{(2^{-k}+\rho(x,z))^{1+\epsilon}} \\
&\quad \times \Big[\frac{2^{-l\epsilon}}{(2^{-l}+\rho(z,y))^{1+\epsilon}} + \frac{2^{-l\epsilon}}{(2^{-l}+\rho(x,y))^{1+\epsilon}}\Big] d\mu(z) \\
&\le C\Big(\frac{\rho(x,x')}{2^{-l}+\rho(x,y)}\Big)^{\sigma} \frac{2^{-k\epsilon}}{(2^{-l}+\rho(x,y))^{1+\epsilon}} \int_{W_1} \frac{2^{-l\epsilon}}{(2^{-l}+\rho(z,y))^{1+\epsilon}} d\mu(z) \\
&\quad + \Big(\frac{\rho(x,x')}{2^{-l}+\rho(x,y)}\Big)^{\sigma} \frac{2^{-l\epsilon}}{(2^{-l}+\rho(x,y))^{1+\epsilon}} \int_{W_1} \frac{2^{-k\epsilon}}{(2^{-k}+\rho(x,z))^{1+\epsilon}} d\mu(z) \\
&\le C\Big(\frac{\rho(x,x')}{2^{-l}+\rho(x,y)}\Big)^{\sigma} \frac{2^{-l\epsilon}}{(2^{-l}+\rho(x,y))^{1+\epsilon}}
\end{aligned}
$$

since $k \ge l$.

Finally, note that if $z \in W_3$, then $\rho(x,z)) \le \frac{1}{2A}(2^{-l}+\rho(x,y))$ and $\rho(x,z) \le 2A\rho(x,x')$. By the estimate (3.15) and (3.14),

$$
\begin{aligned}
III &\le C \int_{W_3} \Big[\frac{2^{-k\epsilon}}{(2^{-k}+\rho(x,z))^{1+\epsilon}} + \frac{2^{-k\epsilon}}{(2^{-k}+\rho(x',z))^{1+\epsilon}}\Big] \\
&\quad \times \Big(\frac{\rho(z,x)}{2^{-l}+\rho(x,y)}\Big)^{\sigma} \frac{2^{-l\epsilon}}{(2^{-l}+\rho(x,y))^{1+\epsilon}} d\mu(z) \\
&\le C\Big(\frac{\rho(x,x')}{2^{-l}+\rho(x,y)}\Big)^{\sigma} \frac{2^{-l\epsilon}}{(2^{-l}+\rho(x,y))^{1+\epsilon}}.
\end{aligned}
$$

This implies (3.22) for the case where $\rho(x,x') \le \frac{1}{4A^2}(2^{-l}+\rho(x,y))$ and hence Lemma 3.9. ∎

Lemma 3.10. *Suppose that $D_k(x,y), k \in \mathbb{Z}$, the kernels of operators D_k, satisfy the same conditions as in Lemma 3.6. Then there exists a constant C which depends only on $\sigma' \le \sigma$ and $\sigma' < \epsilon$ but not on l, such that if $l \ge k$,*

$$
\begin{aligned}
&|D_kD_l(x,y) - D_kD_l(x',y)| \\
&\quad \le C\Big(\frac{\rho(x,x')}{2^{-k}+\rho(x,y)}\Big)^{\sigma'} \frac{2^{-k(\epsilon-\sigma')}}{(2^{-k}+\rho(x,y))^{1+(\epsilon-\sigma')}}
\end{aligned} \tag{3.23}
$$

for $\rho(x,x') \le \frac{1}{2A}(2^{-k}+\rho(x,y))$.

The proof of this lemma is similar to the previous one. Instead we only use the size condition on the kernel of D_l. We leave the details to the reader.

As a consequence of Lemma 3.9 and Lemma 3.10, we have

Lemma 3.11. *Suppose that $\{S_k\}_{k\in\mathbb{Z}}$ is an approximation to the identity . Set $D_k = S_k - S_{k-1}$ for all $k \in \mathbb{Z}$. Then for any $\sigma' \le \sigma$ and $\sigma' < \epsilon$, there exists a constant C which depends only on σ, σ' and ϵ but not on k, l such that*

$$|D_kD_l(x,y) - D_kD_l(x',y)| \\ \leq C\Big(\frac{\rho(x,x')}{2^{-l}+\rho(x,y)}\Big)^{\sigma} \frac{2^{-l\epsilon}}{(2^{-l}+\rho(x,y))^{1+\epsilon}} \tag{3.24}$$

for $k \geq l$ *and* $\rho(x,x') \leq \frac{1}{2A}(2^{-l}+\rho(x,y))$;

$$|D_kD_l(x,y) - D_kD_l(x',y)| \\ \leq C\Big(\frac{\rho(x,x')}{2^{-k}+\rho(x,y)}\Big)^{\sigma'} \frac{2^{-k(\epsilon-\sigma')}}{(2^{-k}+\rho(x,y))^{1+(\epsilon-\sigma')}} \tag{3.25}$$

for $k \leq l$ *and* $\rho(x,x') \leq \frac{1}{2A}(2^{-k}+\rho(x,y))$;

$$|D_kD_l(x,y) - D_kD_l(x,y')| \\ \leq C\Big(\frac{\rho(y,y')}{2^{-k}+\rho(x,y)}\Big)^{\sigma} \frac{2^{-k\epsilon}}{(2^{-k}+\rho(x,y))^{1+\epsilon}} \tag{3.26}$$

for $k \leq l$ *and* $\rho(y,y') \leq \frac{1}{2A}(2^{-k}+\rho(x,y))$;

$$|D_kD_l(x,y) - D_kD_l(x,y')| \\ \leq C\Big(\frac{\rho(y,y')}{2^{-l}+\rho(x,y)}\Big)^{\sigma'} \frac{2^{-l(\epsilon-\sigma')}}{(2^{-l}+\rho(x,y))^{1+(\epsilon-\sigma')}} \tag{3.27}$$

for $l \leq k$ *and* $\rho(y,y') \leq \frac{1}{2A}(2^{-l}+\rho(x,y))$.

We now give the double Lipschitz estimates.

Lemma 3.12. *Suppose that* $\{S_k\}_{k\in\mathbb{Z}}$ *is an approximation to the identity and* $S_k(x,y)$, *the kernels of* S_k, *satisfy the condition (3.19). Set* $D_k = S_k - S_{k-1}$ *for all* $k \in \mathbb{Z}$. *Then for any* $\sigma' \leq \sigma$ *and* $\sigma' < \epsilon$, *there exists a constant* C *which depends only on* σ, σ' *and* ϵ *but not on* k *or* l, *such that*

$$|D_kD_l(x,y) - D_kD_l(x',y) - D_kD_l(x,y') + D_kD_l(x',y')| \\ \leq C\Big(\frac{\rho(x,x')}{2^{-(k\wedge l)}+\rho(x,y)}\Big)^{\sigma'}\Big(\frac{\rho(y,y')}{2^{-(k\wedge l)}+\rho(x,y)}\Big)^{\sigma'} \frac{2^{-(k\wedge l)(\epsilon-\sigma')}}{(2^{-(k\wedge l)}+\rho(x,y))^{1+(\epsilon-\sigma')}} \tag{3.28}$$

for $\rho(x,x') \leq \frac{1}{2A}(2^{-(k\wedge l)}+\rho(x,y))$ *and* $\rho(y,y') \leq \frac{1}{2A}(2^{-(k\wedge l)}+\rho(x,y))$.

Proof. We only prove the case where $k \leq l$. The proof of the case where $k \geq l$ is similar. We also only consider $\rho(x,x') \leq \frac{1}{4A^2}(2^{-(k\wedge l)}+\rho(x,y))$ and $\rho(y,y') \leq \frac{1}{4A^2}(2^{-(k\wedge l)}+\rho(x,y))$. We leave the details for other cases to the reader. Using the moment condition, we write

$$
\begin{aligned}
&|D_kD_l(x,y) - D_kD_l(x',y) - D_kD_l(x,y') + D_kD_l(x',y')| \\
&\quad = \Big| \int [D_k(x,z) - D_k(x',z)][D_l(z,y') - D_l(z,y')] d\mu(z) \Big| \\
&\quad = \Big| \int [D_k(x,z) - D_k(x',z) - D_k(x,y) - D_k(x',y)] \\
&\qquad\qquad \times [D_l(z,y') - D_l(z,y')] d\mu(z) \Big| \\
&\quad \le \int |D_k(x,z) - D_k(x',z) - D_k(x,y) - D_k(x',y)| \\
&\qquad\qquad \times |D_l(z,y') - D_l(z,y')| d\mu(z) \\
&\quad = \int_{W_1} \cdots + \int_{W_2} \cdots + \int_{W_3} \cdots + \int_{W_4} \cdots = I + II + III + IV
\end{aligned}
$$

where

$$
\begin{aligned}
W_1 &= \Big\{ z \in X : \rho(y,y') \le \frac{1}{2A}(2^{-l} + \rho(z,y)) \le \frac{1}{4A^2}(2^{-k} + \rho(x,y)) \Big\}, \\
W_2 &= \Big\{ z \in X : \rho(y,y') \le \frac{1}{4A^2}(2^{-k} + \rho(x,y)) \le \frac{1}{2A}(2^{-l} + \rho(z,y)) \\
&\qquad\qquad \text{and } \rho(x,x') \le \frac{1}{2A}(2^{-k} + \rho(z,x)) \Big\}, \\
W_3 &= \Big\{ z \in X : \rho(y,y') \le \frac{1}{4A^2}(2^{-k} + \rho(x,y)) \le \frac{1}{2A}(2^{-l} + \rho(z,y)) \\
&\qquad\qquad \text{and } \rho(x,x') \ge \frac{1}{2A}(2^{-k} + \rho(z,x)) \Big\}, \\
W_4 &= \Big\{ z \in X : \rho(y,y') > \frac{1}{2A}(2^{-l} + \rho(z,y)) \Big\}.
\end{aligned}
$$

Note that if $z \in W_1$, then $\rho(z,y)) \le \frac{1}{2A}(2^{-k}+\rho(x,y))$ and $\rho(x,x') \le \frac{1}{2A}(2^{-k}+\rho(x,y))$. Using the condition (3.19), we obtain

$$
\begin{aligned}
I &\le C \int_{W_1} \Big(\frac{\rho(z,y)}{2^{-k} + \rho(x,y)} \Big)^{\sigma} \Big(\frac{\rho(x,x')}{2^{-k} + \rho(x,y)} \Big)^{\sigma} \\
&\quad \times \frac{2^{-k\epsilon}}{(2^{-k} + \rho(x,y))^{1+\epsilon}} \Big(\frac{\rho(y,y')}{2^{-l} + \rho(z,y)} \Big)^{\sigma} \frac{2^{-l\epsilon}}{(2^{-l} + \rho(z,y))^{1+\epsilon}} d\mu(z) \\
&\le C \Big(\frac{\rho(x,x')}{2^{-k} + \rho(x,y)} \Big)^{\sigma} \Big(\frac{\rho(y,y')}{2^{-k} + \rho(x,y)} \Big)^{\sigma} \frac{2^{-k\epsilon}}{(2^{-k} + \rho(x,y))^{1+\epsilon}}.
\end{aligned}
$$

If $z \in W_4$, then $2A\rho(y,y') > \rho(z,y), \rho(z,y)) \le \frac{1}{2A}(2^{-k} + \rho(x,y))$ and $\rho(x,x') \le \frac{1}{2A}(2^{-k} + \rho(x,y))$. Thus,

$$IV \le C\int_{W_4}\Big(\frac{\rho(z,y)}{2^{-k}+\rho(x,y)}\Big)^{\sigma}\Big(\frac{\rho(x,x')}{2^{-k}+\rho(x,y)}\Big)^{\sigma}\frac{2^{-k\epsilon}}{(2^{-k}+\rho(x,y))^{1+\epsilon}}$$
$$\times\Big[\frac{2^{-l\epsilon}}{(2^{-l}+\rho(z,y))^{1+\epsilon}}+\frac{2^{-l\epsilon}}{(2^{-l}+\rho(z,y'))^{1+\epsilon}}\Big]d\mu(z)$$
$$\le C\Big(\frac{\rho(x,x')}{2^{-k}+\rho(x,y)}\Big)^{\sigma}\Big(\frac{\rho(y,y')}{2^{-k}+\rho(x,y)}\Big)^{\sigma}\frac{2^{-k\epsilon}}{(2^{-k}+\rho(x,y))^{1+\epsilon}}.$$

For term II, by the smoothness condition,

$$II \le C\int_{W_2}\Big[\Big(\frac{\rho(x,x')}{2^{-k}+\rho(x,z)}\Big)^{\sigma}\frac{2^{-k\epsilon}}{(2^{-k}+\rho(x,z))^{1+\epsilon}}$$
$$+\Big(\frac{\rho(x,x')}{2^{-k}+\rho(x,y)}\Big)^{\sigma}\frac{2^{-k\epsilon}}{(2^{-k}+\rho(x,y))^{1+\epsilon}}\Big]$$
$$\times\Big(\frac{\rho(y,y')}{2^{-l}+\rho(z,y)}\Big)^{\sigma}\frac{2^{-l\epsilon}}{(2^{-l}+\rho(z,y))^{1+\epsilon}}d\mu(z)$$
$$\le C\Big(\frac{\rho(x,x')}{2^{-k}}\Big)^{\sigma'}\Big(\frac{\rho(y,y')}{2^{-k}+\rho(x,y)}\Big)^{\sigma}\frac{2^{-l\epsilon}}{(2^{-k}+\rho(x,y))^{1+\epsilon}}$$
$$\times\int_{W_2}\frac{2^{-k\epsilon}}{(2^{-k}+\rho(x,z))^{1+\epsilon}}d\mu(z)$$
$$+C\Big(\frac{\rho(x,x')}{2^{-k}+\rho(x,y)}\Big)^{\sigma}\Big(\frac{\rho(y,y')}{2^{-k}+\rho(x,y)}\Big)^{\sigma}\frac{2^{-k\epsilon}}{(2^{-k}+\rho(x,y))^{1+\epsilon}}$$
$$\le C\Big(\frac{\rho(x,x')}{2^{-k}+\rho(x,y)}\Big)^{\sigma'}\Big(\frac{\rho(y,y')}{2^{-k}+\rho(x,y)}\Big)^{\sigma}\frac{2^{-k(\epsilon-\sigma')}}{(2^{-k}+\rho(x,y))^{1+(\epsilon-\sigma')}}.$$

Finally, if $z \in W_3$, then $2A\rho(x,x')2^{k} \ge 1$. Using the size and smoothness conditions,

$$III \le C\int_{W_3}\Big[\frac{2^{-k\epsilon}}{(2^{-k}+\rho(x,z))^{1+\epsilon}}+\frac{2^{-k\epsilon}}{(2^{-k}+\rho(x',z))^{1+\epsilon}}+(\frac{\rho(x,x')}{2^{-k}+\rho(x,y)})^{\sigma}$$
$$\cdot\frac{2^{-k\epsilon}}{(2^{-k}+\rho(x,y))^{1+\epsilon}}\Big]\times(\frac{\rho(y,y')}{2^{-l}+\rho(z,y)})^{\sigma}\frac{2^{-l\epsilon}}{(2^{-l}+\rho(z,y))^{1+\epsilon}}d\mu(z)$$
$$\le C\Big(\frac{\rho(y,y')}{2^{-k}+\rho(z,y)}\Big)^{\sigma}\frac{2^{-k\epsilon}}{(2^{-k}+\rho(z,y))^{1+\epsilon}}$$
$$+C\Big(\frac{\rho(x,x')}{2^{-k}+\rho(x,y)}\Big)^{\sigma}\Big(\frac{\rho(y,y')}{2^{-k}+\rho(x,y)}\Big)^{\sigma}\frac{2^{-k\epsilon}}{(2^{-k}+\rho(x,y))^{1+\epsilon}}$$
$$\le C\Big(\frac{\rho(x,x')}{2^{-k}+\rho(x,y)}\Big)^{\sigma'}\Big(\frac{\rho(y,y')}{2^{-k}+\rho(x,y)}\Big)^{\sigma}\frac{2^{-k(\epsilon-\sigma')}}{(2^{-k}+\rho(x,y))^{1+(\epsilon-\sigma')}}$$

which implies (3.28) for the case where $k \le l$ and hence, Lemma 3.12. ■

It is easy to check that the approximation to the identity constructed by Coifman's idea satisfies the condition (3.19). The following result implies that S_kS_k, where S_k is an approximation to the identity , satisfies the condition (3.19).

Lemma 3.13. *Suppose that $\{S_k\}_{k\in\mathbb{Z}}$ is an approximation to the identity . Then S_kS_k is an approximation to the identity and $S_kS_k(x,y)$, the kernels of S_kS_k, satisfy the following conditions: For any $\sigma'\le\sigma$ and $\sigma'<\epsilon$, there exists a constant C such that*

$$|S_kS_k(x,y)|\le C\frac{2^{-k\epsilon}}{(2^{-k}+\rho(x,y))^{1+\epsilon}};$$

$$|S_kS_k(x,y)-S_kS_k(x',y)|\le C\Big(\frac{\rho(x,x')}{2^{-k}+\rho(x,y)}\Big)^{\sigma'}\frac{2^{-k(\epsilon-\sigma')}}{(2^{-k}+\rho(x,y))^{1+(\epsilon-\sigma')}}$$

for $\rho(x,x')\le\frac{1}{2A}(2^{-k}+\rho(x,y)$;

$$|S_kS_k(x,y)-S_kS_k(x,y')|\le C\Big(\frac{\rho(y,y')}{2^{-k}+\rho(x,y)}\Big)^{\sigma'}\frac{2^{-k(\epsilon-\sigma')}}{(2^{-k}+\rho(x,y))^{1+(\epsilon-\sigma')}}$$

for $\rho(y,y')\le\frac{1}{2A}(2^{-k}+\rho(x,y)$;

$$\int S_kS_k(x,y)d\mu(y)=1$$

for all $x\in X$;

$$\int S_kS_k(x,y)d\mu(x)=1$$

for all $y\in X$;

$$\begin{aligned}&|S_kS_k(x,y)-S_kS_k(x',y)-S_kS_k(x,y')+S_kS_k(x',y')|\\&\quad\le C\Big(\frac{\rho(x,x')}{2^{-k}+\rho(x,y)}\Big)^{\sigma'}\Big(\frac{\rho(y,y')}{2^{-k}+\rho(x,y)}\Big)^{\sigma'}\frac{2^{-k(\epsilon-\sigma')}}{(2^{-k}+\rho(x,y))^{1+(\epsilon-\sigma')}}\end{aligned}$$

for $\rho(x,x')\le\frac{1}{2A}(2^{-k}+\rho(x,y))$ *and* $\rho(y,y')\le\frac{1}{2A}(2^{-k}+\rho(x,y))$.

Note that when $k=l$, the moment condition is not needed. So the proof of Lemma 3.13 is similar to the proofs of the above lemmas. We leave the details to the reader.

As in the case of $\mathbb{R}^n$, we define "father" and "mother" functions on $X\times X$ where X is a space of homogeneous type.

Definition 3.14. A family of functions $\{S_k(x,y)\}_{k\in\mathbb{Z}}$ is said to be a family of "father functions" if $\{S_k(x,y)\}_{k\in\mathbb{Z}}$ are kernels of an approximation to the identity, which satisfies the double Lipschitz condition (3.19) with $\sigma=\epsilon$. A family of functions $\{D_k(x,y)\}_{k\in\mathbb{Z}}$ is said to be a family of "mother functions" if $D_k(x,y)=S_k(x,y)-S_{k-1}(x,y)$ where $\{S_k(x,y)\}_{k\in\mathbb{Z}}$ are father functions.

We are now ready to prove Calderón's identity on spaces of homogeneous type in the next section.

3.4 Calderón's Identity on Spaces of Homogeneous Type

The purpose of this section is to prove Calderón's identity on spaces of homogeneous type. We first prove the following fundamental result which is a substitute for the missing Fourier transformation.

Theorem 3.15. *([HS] and [H1]) Suppose that $\{D_k(x,y)\}_{k\in\mathbb{Z}}$ is a family of mother functions. Then there exist families of operators $\{\widetilde{D}_k\}_{k\in\mathbb{Z}}$ and $\{\widetilde{\widetilde{D}}_k\}_{k\in\mathbb{Z}}$ such that for $f\in\mathcal{M}_0(\beta,\gamma)$,*

$$f=\sum_k \widetilde{D}_k D_k(f)=\sum_k D_k\widetilde{\widetilde{D}}_k(f) \tag{3.29}$$

where the series converge in the norm of $\mathcal{M}(\beta',\gamma')$ for $0<\beta'<\beta$ and $0<\gamma'<\gamma$, and in the $L^p(X)$ spaces for $1<p<\infty$. Moreover, $\widetilde{D}_k(x,y)$, the kernels of $\widetilde{D}_k$, as functions of x are wavelets (in the sense given in the Introduction) satisfying (3.14) and (3.15) with ϵ replaced by ϵ', for $0<\epsilon'<\epsilon$, and $\int\widetilde{D}_k(x,y)d\mu(y)=\int\widetilde{D}_k(x,y)d\mu(x)=0$ for all $k\in\mathbb{Z}$, and $\widetilde{\widetilde{D}}_k(x,y)$, the kernels of $\widetilde{\widetilde{D}}_k$, as functions of y are wavelets satisfying the conditions (3.14) and (3.16) with ϵ replaced by ϵ', for $0<\epsilon'<\epsilon$, and $\int\widetilde{\widetilde{D}}_k(x,y)d\mu(y)=\int\widetilde{\widetilde{D}}_k(x,y)d\mu(x)=0$ for all $k\in\mathbb{Z}$.

Before proving the above theorem, we would like to point out that in the standard $\mathbb{R}^n$ case, we begin with a function $\psi(x)\in L^1(\mathbb{R}^n)$ whose integral over $\mathbb{R}^n$ is zero and whose Fourier transform $\hat{\psi}(\xi),\xi\in\mathbb{R}^n$, satisfies $\int_0^\infty|\hat{\psi}(t\xi)|^2\frac{dt}{t}=1$ for every $\xi\neq0$. To construct a function satisfying this condition, we begin with a function $\psi(x)$ which is sufficiently regular, localized, the integral over $\mathbb{R}^n$ being zero and which is a radial function. Then there is a constant $c>0$ such that $c\psi(x)$ satisfies this condition. A. Grossmann and J. Morlet ([GrMo]) defined the "wavelet coefficients" of $f\in L^2(\mathbb{R}^n)$ by

$$\alpha(u,t)=(f,\psi_{(u,t)}) \tag{3.30}$$

where $\psi_{(u,t)}=\psi_t(x-u)$ and $\psi_t=t^{-n}\psi(\frac{x}{t})$.

We then can construct f from the coefficients by

$$f(x)=\int_0^\infty\int_{\mathbb{R}^n}\alpha(u,t)\psi_{(u,t)}du\frac{dt}{t}. \tag{3.31}$$

An equivalent formulation of the above equality is Calderón's identity given by

$$I = \int_0^\infty Q_t Q_t^* \frac{dt}{t} \tag{3.32}$$

where $Q_t(f) = f * \psi_t$ and Q_t^* is the adjoint of Q_t.

It is clear that Theorem 3.15 provides a Calderón's identity on spaces of homogeneous type. Therefore Theorem 3.15 also provides a continuous version of wavelet expansions on spaces of homogeneous type. Moreover, Theorem 3.15 says that such a continuous version of wavelet expansions holds not only on L^2, but also on L^p for all $p : 1 < p < \infty$.

We now prove Theorem 3.15. First of all, we show the following result.

Proposition 3.16 *Suppose that $\{S_k\}_{k\in\mathbb{Z}}$ is an approximation to the identity and its kernels satisfy the condition (3.19) with $\sigma = \epsilon \le \theta$. Set $D_k = S_k - S_{k-1}$ for all $k \in \mathbb{Z}$. Let $T_N = \sum_k D_k^N D_k$ where $D_k^N = \sum_{|j|\le N} D_{k+j}$ and N is a fixed positive integer. Then T_N^{-1} exists if N is a sufficiently large integer. Moreover, there exists a constant C such that for $f \in \mathcal{M}_0(x_1, r, \beta, \gamma)$ with $x_1 \in X, r > 0$ and $0 < \beta, \gamma < \epsilon$, then $T_N^{-1}(f) \in \mathcal{M}_0(x_1, r, \beta, \gamma)$ and*

$$\|T_N^{-1}(f)\|_{\mathcal{M}(x_1,r,\beta,\gamma)} \le C\|f\|_{\mathcal{M}(x_1,r,\beta,\gamma)}. \tag{3.33}$$

By Coifman's decomposition of identity operator as described in Chapter 1,

$$I = \sum_k \sum_l D_k D_l = T_N + R_N \tag{3.34}$$

where $R_N = \sum_{|l|>N} \sum_k D_{k+l} D_k$.

It suffices to show that R_N satisfies all conditions of Theorem 2.4 with the constant C in (2.3) less than 1 for a fixed large positive integer N. The following lemma gives the required estimates on R_N.

Lemma 3.17. *Suppose that $\{S_k\}_{k\in\mathbb{Z}}$ is an approximation to the identity and its kernels satisfy the condition (3.19) with $\sigma = \epsilon \le \theta$. Set $D_k = S_k - S_{k-1}$ for all $k \in \mathbb{Z}$. Let $R_N = \sum_{|l|>N} \sum_k D_{k+l} D_k$. Then for $0 < \epsilon' < \epsilon, R_N \in CZK(\epsilon') \cap SWBP, R_N(1) = (R_N)^*(1) = 0$. Moreover, $R_N(x, y)$, the kernels of R_N satisfy the following estimates: there exist constants $0 < C < \infty$, which is independent of N, and $\delta > 0$ such that*

$$|R_N(x, y)| \le C2^{-N\delta} \rho(x, y)^{-1}; \tag{3.35}$$

$$|R_N(x,y) - R_N(x',y)| \le C2^{-N\delta}\rho(x,x')^{\epsilon'}\rho(x,y)^{-(1+\epsilon')} \tag{3.36}$$

for $\rho(x,x') \le \frac{1}{2A}\rho(x,y)$;

$$|R_N(x,y) - R_N(x,y')| \le C2^{-N\delta}\rho(y,y')^{\epsilon'}\rho(x,y)^{-(1+\epsilon')} \tag{3.37}$$

for $\rho(y,y') \le \frac{1}{2A}\rho(x,y)$;

$$\begin{aligned} &|R_N(x,y) - R_N(x',y) - R_N(x,y') + R_N(x',y')| \\ &\quad \le C2^{-N\delta}\rho(x,x')^{\epsilon'}\rho(y,y')^{\epsilon'}\rho(x,y)^{-(1+2\epsilon')} \end{aligned} \tag{3.38}$$

for $\rho(x,x') \le \frac{1}{2A}\rho(x,y)$ *and* $\rho(y,y') \le \frac{1}{2A}\rho(x,y)$;

$$|\langle R_N, f\rangle| \le C2^{-N\delta}r \tag{3.39}$$

for all $f \in C_0^\eta(X\times X)$ *with* $\mathrm{supp}(f) \subseteq B(x_1,r)\times B(y_1,r), x_1$ *and* $y_1 \in X, \|f\|_\infty \le 1, \|f(.,y)\|_\eta \le r^{-\eta}$, *and* $\|f(x,)\|_\eta \le r^{-\eta}$ *for all* x *and* $y \in X$.

Assuming Lemma 3.17 for the moment, by Theorem 2.4, for all $f \in \mathcal{M}_0(x_1,r,\beta,\gamma)$ with $x_1 \in X, r > 0$ and $0 < \beta,\gamma < \epsilon$, there exists a constant C such that

$$\|R_N(f)\|_{\mathcal{M}(x_1,r,\beta,\gamma)} \le C2^{-N\delta}\|f\|_{\mathcal{M}(x_1,r,\beta,\gamma)}$$

which, together with the fact that $T_N^{-1} = \sum\limits_{k=0}(R_N)^k$ if N is large enough, implies (3.33) and hence Proposition 3.16.

We now prove Lemma 3.17. We write

$$|R_N(x,y)| = \Big|\sum_{|l|>N}\sum_k D_{k+l}D_k(x,y)\Big| \le \sum_{|l|>N}\sum_k |D_{k+l}D_k(x,y)|.$$

Using the estimates in Lemma 3.7 with $\sigma = \epsilon$, we have

$$|R_N(x,y)| \le C\sum_{|l|>N}\sum_k 2^{-|l|\epsilon'}\frac{2^{-((k+l)\wedge k)\epsilon}}{(2^{-((k+l)\wedge k)} + \rho(x,y))^{1+\epsilon}} \le C2^{-N\epsilon'}\rho(x,y)^{-1}$$

which yields (3.35).

To see (3.36), by the geometric mean of estimates in Lemma 3.7 and Lemma 3.11, for $\rho(x,x') \le \frac{1}{2A}\rho(x,y)$ we get

$$\begin{aligned} &|D_{k+l}D_k(x,y) - D_{k+l}D_k(x',y)| \\ &\quad \le C2^{-|l|\delta}\Big(\frac{\rho(x,x')}{2^{-((k+l)\wedge k)} + \rho(x,y)}\Big)^{\epsilon'}\frac{2^{-((k+l)\wedge k)\epsilon'}}{(2^{-((k+l)\wedge k)} + \rho(x,y))^{1+\epsilon'}} \end{aligned} \tag{3.40}$$

where $0 < \epsilon' < \epsilon$ and $\delta > 0$. Thus, for $\rho(x,x') \le \frac{1}{2A}\rho(x,y)$,

$$\begin{aligned}
&|R_N(x,y) - R_N(x',y)| \\
&\quad = \Big|\sum_{|l|>N}\sum_k [D_{k+l}D_k(x,y) - D_{k+l}D_k(x',y)]\Big| \\
&\quad \le \sum_{|l|>N}\sum_k |D_{k+l}D_k(x,y) - D_{k+l}D_k(x',y)| \\
&\quad \le C\sum_{|l|>N}\sum_k 2^{-|l|\delta}\Big(\frac{\rho(x,x')}{2^{-(k\wedge l)}+\rho(x,y)}\Big)^{\epsilon'}\frac{2^{-(k\wedge l)\epsilon'}}{(2^{-(k\wedge l)}+\rho(x,y))^{1+\epsilon'}} \\
&\quad \le C2^{-N\delta}\rho(x,x')^{\epsilon'}\rho(x,y)^{-(1+\epsilon')}.
\end{aligned}$$

The proof of (3.37) is the same. By the geometric mean of estimates in Lemma 3.7 and Lemma 3.11, we obtain that for $\rho(y,y') \le \frac{1}{2A}\rho(x,y)$,

$$\begin{aligned}
&|D_{k+l}D_k(x,y) - D_{k+l}D_k(x,y')| \\
&\quad \le C2^{-|l|\delta}\Big(\frac{\rho(y,y')}{2^{-((k+l)\wedge k)}+\rho(x,y)}\Big)^{\epsilon'}\frac{2^{-((k+l)\wedge k)\epsilon'}}{(2^{-((k+l)\wedge k)}+\rho(x,y))^{1+\epsilon'}}
\end{aligned} \tag{3.41}$$

where $0<\epsilon'<\epsilon$ and $\delta>0$. Thus, for $\rho(y,y') \le \frac{1}{2A}\rho(x,y)$,

$$\begin{aligned}
&|R_N(x,y) - R_N(x,y')| \\
&\quad = \Big|\sum_{|l|>N}\sum_k [D_{k+l}D_k(x,y) - D_{k+l}D_k(x,y')\Big| \\
&\quad \le \sum_{|l|>N}\sum_k |D_{k+l}D_k(x,y) - D_{k+l}D_k(x,y')| \\
&\quad \le C\sum_{|l|>N}\sum_k 2^{-|l|\delta}\Big(\frac{\rho(y,y')}{2^{-(k\wedge l)}+\rho(x,y)}\Big)^{\epsilon'}\frac{2^{-(k\wedge l)\epsilon'}}{(2^{-(k\wedge l)}+\rho(x,y))^{1+\epsilon'}} \\
&\quad \le C2^{-N\delta}\rho(y,y')^{\epsilon'}\rho(x,y)^{-(1+\epsilon')}.
\end{aligned}$$

To show (3.38), by the geometric mean of the estimates in Lemma 3.12, (3.40) and (3.41), we get that for $\rho(x,x') \le \frac{1}{2A}\rho(x,y)$ and $\rho(y,y') \le \frac{1}{2A}\rho(x,y)$,

$$\begin{aligned}
&|D_{k+l}D_k(x,y) - D_{k+l}D_k(x',y) - D_{k+l}D_k(x,y') - D_{k+l}D_k(x',y')| \\
&\le C2^{-|l|\delta}\Big(\frac{\rho(x,x')}{2^{-((k+l)\wedge k)}+\rho(x,y)}\Big)^{\epsilon'}\Big(\frac{\rho(y,y')}{2^{-((k+l)\wedge k)}+\rho(x,y)}\Big)^{\epsilon'} \\
&\qquad \times\frac{2^{-((k+l)\wedge k)(\epsilon-\epsilon')}}{(2^{-((k+l)\wedge k)}+\rho(x,y))^{1+(\epsilon-\epsilon')}}.
\end{aligned}$$

Thus, for $\rho(x,x') \le \frac{1}{2A}\rho(x,y)$ and $\rho(y,y') \le \frac{1}{2A}\rho(x,y)$,

$$
\begin{aligned}
&|R_N(x,y) - R_N(x',y) - R_N(x,y') + R_N(x',y')| \\
&\quad \le \sum_{|l|>N} \sum_k |D_{k+l}D_k(x,y) - D_{k+l}D_k(x',y) - D_{k+l}D_k(x,y') \\
&\qquad\qquad - D_{k+l}D_k(x',y')| \\
&\quad \le C \sum_{|l|>N} \sum_k 2^{-|l|\delta} \Big(\frac{\rho(x,x')}{2^{-((k+l)\wedge k)} + \rho(x,y)} \Big)^{\epsilon'} \Big(\frac{\rho(y,y')}{2^{-((k+l)\wedge k)} + \rho(x,y)} \Big)^{\epsilon'} \\
&\qquad\qquad \times \frac{2^{-((k+l)\wedge k)(\epsilon-\epsilon')}}{(2^{-((k+l)\wedge k)} + \rho(x,y))^{1+(\epsilon-\epsilon')}} \\
&\quad \le C 2^{-N\delta} \rho(x,x')^{\epsilon'} \rho(y,y')^{\epsilon'} \rho(x,y)^{-(1+2\epsilon')}.
\end{aligned}
$$

Finally, we prove (3.39). Suppose that $f \in C_0^\eta(X \times X)$ with $\mathrm{supp}(f) \subseteq B(x_1,r) \times B(y_1,r), x_1$ and $y_1 \in X, \|f\|_\infty \le 1, \|f(\cdot,y)\|_\eta \le r^{-\eta}$, and $\|f(x,\cdot)\|_\eta \le r^{-\eta}$ for all x and $y \in X$. We get

$$
\begin{aligned}
|\langle D_{k+l}D_k, f\rangle| &= \left| \int\int D_{k+l}D_k(x,y) f(x,y) d\mu(y) d\mu(x) \right| \\
&\le C 2^{-|l|\epsilon'} \|f\|_\infty r \le C 2^{-|l|\epsilon'} r.
\end{aligned} \tag{3.42}
$$

On the other hand, for $\eta < \epsilon$,

$$
\begin{aligned}
&|\langle D_{k+l}D_k, f\rangle| \\
&\quad \le \left| \int\int\int D_{k+l}(x,z) D_k(z,y) f(x,y) d\mu(z) d\mu(y) d\mu(x) \right| \\
&\quad = \left| \int\int\int D_{k+l}(x,z) D_k(z,y) [f(x,y) - f(x,z)] d\mu(z) d\mu(y) d\mu(x) \right| \\
&\quad \le C 2^{-k\eta} r^{-\eta} r.
\end{aligned} \tag{3.43}
$$

We also have

$$
\begin{aligned}
|\langle D_{k+l}D_k, f\rangle| &\le \left| \int\int\int D_{k+l}(x,z) D_k(z,y) f(x,y) d\mu(z) d\mu(y) d\mu(x) \right| \\
&\le C 2^k r^2.
\end{aligned} \tag{3.44}
$$

The geometric means of (3.42) and (3.43) yields

$$
|\langle D_{k+l}D_k, f\rangle| \le C 2^{-|l|\delta} 2^{-k\eta'} r^{-\eta'} r, \tag{3.45}
$$

and the geometric means of (3.42) and (3.44) implies

$$
|\langle D_{k+l}D_k, f\rangle| \le C 2^{-|l|\delta} 2^{k\eta''} r^{\eta''} r \tag{3.46}
$$

where δ, η' and $\eta'' > 0$. Therefore,

$$
\begin{aligned}
|\langle R_N, f\rangle| &= \Big|\Big\langle \sum_{|l|>N}\sum_k D_{k+l}D_k, f\Big\rangle\Big| \\
&\leq \sum_{|l|>N}\sum_{2^{-k}>r} |\langle D_{k+l}D_k, f\rangle| + \sum_{|l|>N}\sum_{2^{-k}\leq r} |\langle D_{k+l}D_k, f\rangle| \\
&\leq C2^{-N\delta}r
\end{aligned}
$$

which implies (3.39), and hence Lemma 3.17.

We now return to the proof of Theorem 3.15. Let $\widetilde{D}_k = T_N^{-1}D_k^N$, where D_k^N is defined in Proposition 3.16 and N is a fixed large integer so that T_N^{-1} maps $\mathcal{M}_0(x_1, r, \beta, \gamma)$, for all $x_1 \in X, r > 0$, and $0 < \beta, \gamma < \epsilon$, into $\mathcal{M}_0(x_1, r, \beta, \gamma)$ by Proposition 3.16. It is easy to check that $D_k^N(x, y)$, the kernel of D_k^N, is in $\mathcal{M}_0(y, 2^{-k}, \epsilon, \epsilon)$. Thus, $\widetilde{D}_k(x, y) = T_N^{-1}[D_k^N(\cdot, y)](x)$, the kernel of $\widetilde{D}_k$, is in $\mathcal{M}_0(y, 2^{-k}, \epsilon', \epsilon')$ with $0 < \epsilon' < \epsilon$ by Proposition 3.16. This implies that $\widetilde{D}_k(x, y)$ satisfy the estimates of (3.14) and (3.15) with ϵ replaced by $\epsilon', 0 < \epsilon' < \epsilon$. $\int \widetilde{D}_k(x, y)d\mu(y) = \int \widetilde{D}_k(x, y)d\mu(x) = 0$ for all $k \in \mathbb{Z}$, follow from the facts that $D_k^N(1) = (T_N^{-1})^*(1) = 0$. Similarly, let $\widetilde{\widetilde{D}}_k = D_k^N T_N^{-1}$. Thus, $\widetilde{\widetilde{D}}_k(x, y) = [D_k^N(x, \cdot)T_N^{-1}](y)$, the kernel of $\widetilde{\widetilde{D}}_k$, is in $\mathcal{M}_0(x, 2^{-k}, \epsilon', \epsilon')$ with $0 < \epsilon' < \epsilon$ by Proposition 3.16. This implies that $\widetilde{\widetilde{D}}_k(x, y)$ satisfy the estimates of (3.14) and (3.16) with ϵ replaced by $\epsilon', 0 < \epsilon' < \epsilon$. $\int \widetilde{\widetilde{D}}_k(x, y)d\mu(y) = \int \widetilde{\widetilde{D}}_k(x, y)d\mu(x) = 0$ for all $k \in \mathbb{Z}$, follow from the facts that $(D_k^N)^*(1) = T_N^{-1}(1) = 0$.

All we need to do now is to prove that the series in (3.29) converge in the norm of $\mathcal{M}(\beta', \gamma')$ for $0 < \beta' < \beta$ and $0 < \gamma' < \gamma$, and $L^p(X)$, for $1 < p < \infty$.

Suppose first that $f \in \mathcal{M}_0(\beta, \gamma)$. Then the convergence of (3.29) in $\mathcal{M}(\beta', \gamma')$ is equivalent to

$$
\lim_{M\to\infty} \Big\| \sum_{|k|\leq M} \widetilde{D}_k D_k(f) - f \Big\|_{\mathcal{M}(\beta', \gamma')} = 0
$$

for $0 < \beta' < \beta$ and $0 < \gamma' < \gamma$.

Since

$$
\begin{aligned}
\sum_{|k|\leq M} \widetilde{D}_k D_k(f) &= T_N^{-1}\Big(\sum_{|k|\leq M} D_k^N D_k(f)\Big) = T_N^{-1}\Big(T_N - \sum_{|k|>M} D_k^N D_k(f)\Big) \\
&= f - \lim_{m\to\infty} R_N^m(f) - T_N^{-1}\Big(\sum_{|k|>M} D_k^N D_k(f)\Big),
\end{aligned}
$$

to show the convergence of (3.29) in $\mathcal{M}(\beta', \gamma')$, it suffices to prove

$$
\lim_{m\to\infty} \|R_N^m(f)\|_{\mathcal{M}(\beta', \gamma')} = 0, \tag{3.47}
$$

and

$$\lim_{M\to\infty}\Big\|T_N^{-1}\Big(\sum_{|k|>M} D_k^N D_k(f)\Big)\Big\|_{\mathcal{M}(\beta',\gamma')}=0. \tag{3.48}$$

By Lemma 3.17 and Theorem 2.4, since $0<\beta'<\beta$ and $0<\gamma'<\gamma$,

$$\|R_N^m(f)\|_{\mathcal{M}(\beta',\gamma')}\le (C2^{-N\delta})^m\|f\|_{\mathcal{M}(\beta',\gamma')}\le (C2^{-N\delta})^m\|f\|_{\mathcal{M}(\beta,\gamma)}$$

which implies (3.47). The proof of (3.48) is based on the following estimate

$$\Big\|\sum_{|k|>M} D_k^N D_k(f)\Big\|_{\mathcal{M}(\beta',\gamma')}\le C2^{-\sigma M}\|f\|_{\mathcal{M}(\beta,\gamma)} \tag{3.49}$$

for all $0<\beta'<\beta$ and $0<\gamma'<\gamma$ and some $\sigma>0$, and constant C is independent of f and M.

Assuming (3.49) for the moment, by Proposition 3.16, for $0<\beta'<\beta$ and $0<\gamma'<\gamma$,

$$\begin{aligned}\Big\|T_N^{-1}\Big(\sum_{|k|>M} D_k^N D_k(f)\Big)\Big\|_{\mathcal{M}(\beta',\gamma')}&\le C\Big\|\sum_{|k|>M} D_k^N D_k(f)\Big\|_{\mathcal{M}(\beta',\gamma')}\\&\le C2^{-\sigma M}\|f\|_{\mathcal{M}(\beta,\gamma)}\end{aligned}$$

which yields (3.48).

To prove (3.49), it suffices to show that for $0<\beta'<\beta$ and $0<\gamma'<\gamma$ there exist a constant C which is independent of f and M, and some $\sigma'>0$ such that

$$\Big|\sum_{|k|>M} D_k^N D_k(f)(x)\Big|\le C2^{-\sigma' M}(1+\rho(x,x_0))^{-(1+\gamma')}\|f\|_{\mathcal{M}(\beta,\gamma)}, \tag{3.50}$$

and

$$\begin{aligned}&\Big|\sum_{|k|>M} D_k^N D_k(f)(x)-\sum_{|k|>M} D_k^N D_k(f)(x')\Big| \\&\quad\le C\Big(\frac{\rho(x,x')}{1+\rho(x,x_0)}\Big)^{\beta''}(1+\rho(x,x_0))^{-(1+\gamma')}\|f\|_{\mathcal{M}(\beta,\gamma)}\end{aligned} \tag{3.51}$$

for $\rho(x,x')\le\frac{1}{2}(1+\rho(x,x_0))$ and any $0<\beta'<\beta''<\beta$.

To see this, by taking the geometric mean between (3.51) and the following estimate

$$\left|\sum_{|k|>M} D_k^N D_k(f)(x) - \sum_{|k|>M} D_k^N D_k(f)(x')\right| \tag{3.52}$$
$$\leq \left|\sum_{|k|>M} D_k^N D_k(f)(x)\right| + \left|\sum_{|k|>M} D_k^N D_k(f)(x')\right|$$
$$\leq C2^{-\sigma' M}(1+\rho(x,x_0))^{-(1+\gamma')}\|f\|_{\mathcal{M}(\beta,\gamma)}$$

for $\rho(x,x') \leq \frac{1}{2}(1+\rho(x,x_0))$, we obtain

$$\left|\sum_{|k|>M} D_k^N D_k(f)(x) - \sum_{|k|>M} D_k^N D_k(f)(x')\right| \tag{3.53}$$
$$\leq C2^{-M\sigma}\Big(\frac{\rho(x,x')}{1+\rho(x,x_0)}\Big)^{\beta'}(1+\rho(x,x_0))^{-(1+\gamma')}\|f\|_{\mathcal{M}(\beta,\gamma)}$$

for $\rho(x,x') \leq \frac{1}{2}(1+\rho(x,x_0))$.

Now (3.50) and (3.53), together with the facts that

$$\int \sum_{|k|>M} D_k^N D_k(f)(x)d\mu(x) = \int \sum_{|k|>M} D_k(f)(y)(D_k^N)^*(1)(y)d\mu(y) = 0,$$

imply that

$$\sum_{|k|>M} D_k^N D_k(f)(x)d\mu(x) \in \mathcal{M}_0(\beta',\gamma')$$

and

$$\Big\|\sum_{|k|>M} D_k^N D_k(f)\Big\|_{\mathcal{M}(\beta',\gamma')} \leq C2^{-\sigma' M}\|f\|_{\mathcal{M}(\beta,\gamma)}$$

which gives (3.49).

We now prove (3.50). Denote $D_k^N D_k = E_k$. By Lemma 3.7 and Lemma 3.11, it is easy to check that $E_k(x,y)$, the kernel of E_k, satisfies the estimates (3.29), (3.15) and (3.16) with ϵ replaced by $\epsilon', 0 < \epsilon' < \epsilon, 0 < \beta, \gamma < \epsilon'$, and $E_k(1) = 0$ for all $k \in Z$. We get

$$\left|\sum_{|k|>M} D_k^N D_k(f)(x)\right| = \left|\sum_{|k|>M} E_k(f)(x)\right|$$
$$\leq \left|\int \sum_{k>M} E_k(x,y)[(f)(y) - f(x)]d\mu(y)\right| + \left|\int \sum_{k<-M} E_k(x,y)(f)(y)d\mu(y)\right|$$
$$:= I + II.$$

For term I, we first decompose X into the regions $\rho(x,y) \leq \frac{1}{2A}(1+\rho(x,x_0))$ and $\rho(x,y) > \frac{1}{2A}(1+\rho(x,x_0))$. Using the smoothness condition and size condition on these two regions, respectively, yields

$$\begin{aligned}
I &\leq \sum_{k>M} \int_{\rho(x,y)\leq \frac{1}{2A}(1+\rho(x,x_0))} |E_k(x,y)||f(y)-f(x)|d\mu(y) \\
&\quad + \sum_{k>M} \int_{\rho(x,y)> \frac{1}{2A}(1+\rho(x,x_0))} |E_k(x,y)|[|f(y)|+|f(x)|]d\mu(y) \\
&\leq C \sum_{k>M} \int_{\rho(x,y)\leq \frac{1}{2A}(1+\rho(x,x_0))} |E_k(x,y)|(\frac{\rho(x,y)}{1+\rho(x,x_0)})^{\beta} \\
&\quad \times (1+\rho(x,x_0))^{-(1+\gamma)} d\mu(y) \|f\|_{\mathcal{M}(\beta',\gamma')} \\
&\quad + C \sum_{k>M} \int_{\rho(x,y)> \frac{1}{2A}(1+\rho(x,x_0))} |E_k(x,y)|[(1+\rho(y,x_0))^{-(1+\gamma)} \\
&\quad + (1+\rho(x,x_0))^{-(1+\gamma)} d\mu(y) \|f\|_{\mathcal{M}(\beta',\gamma')} \\
&\leq C \sum_{k>M} \{2^{-k\beta}(1+\rho(x,x_0))^{-(1+\gamma)} + 2^{-k\epsilon'}(1+\rho(x,x_0))^{-(1+\epsilon')} \\
&\quad + 2^{-k\epsilon'}(1+\rho(x,x_0))^{-(1+\gamma)}\}\|f\|_{\mathcal{M}(\beta,\gamma)} \\
&\leq C 2^{-\beta M}(1+\rho(x,x_0))^{-(1+\gamma)}\|f\|_{\mathcal{M}(\beta,\gamma)}.
\end{aligned}$$

For term II, by use of the fact that $\int f(x)d\mu(x) = 0$, we obtain

$$\begin{aligned}
|D_k^N D_k(f)(x)| &= |E_k(f)(x)| = \left| \int [E_k(x,y) - E_k(x,x_0)] f(y) d\mu(y) \right| \\
&\leq \int_{W_1} |E_k(x,y) - E_k(x,x_0)||f(y)|d\mu(y) \\
&\quad + \int_{W_2} |E_k(x,y) - E_k(x,x_0)||f(y)|d\mu(y) \\
&:= II_1 + II_2
\end{aligned}$$

where

$$W_1 = \left\{ z \in X : \rho(y,x_0) \leq \frac{1}{2A}(2^{-k} + \rho(x,x_0)) \right\},$$

and

$$W_2 = \left\{ z \in X : \frac{1}{2A}(2^{-k} + \rho(x,x_0)) < \rho(y,x_0) \right\}.$$

By the smoothness condition on E_k,

$$
\begin{aligned}
II_1 &\le C\int_{W_1}\Big(\frac{\rho(y,x_0)}{2^{-k}+\rho(x,x_0)}\Big)^{\epsilon'}\frac{2^{-k\epsilon'}}{(2^{-k}+\rho(x,x_0))^{1+\epsilon'}}|f(y)|d\mu(y)\\
&\le C\int_{W_1}\Big(\frac{\rho(y,x_0)}{2^{-k}+\rho(x,x_0)}\Big)^{\gamma''}\frac{2^{-k\epsilon'}}{(2^{-k}+\rho(x,x_0))^{1+\epsilon'}}|f(y)|d\mu(y)\\
&\le C\int_{W_1}\Big(\frac{\rho(y,x_0)}{2^{-k}+\rho(x,x_0)}\Big)^{\gamma''}\frac{2^{-k\epsilon'}}{(2^{-k}+\rho(x,x_0))^{1+\epsilon'}}\frac{1}{(1+\rho(y,x_0))^{1+\gamma}}\\
&\qquad\times d\mu(y)\|f\|_{\mathcal{M}(\beta,\gamma)}\\
&\le C2^{\sigma_1 k}(1+\rho(x,x_0))^{-(1+\gamma')}\|f\|_{\mathcal{M}(\beta,\gamma)}
\end{aligned}
$$

where $\gamma' < \gamma''$ and $\sigma_1 = \gamma'' - \gamma' > 0$.

For term II_2, using the size condition on E_k,

$$
\begin{aligned}
II_2 &\le C\int_{W_2}\Big[\frac{2^{-k\epsilon'}}{(2^{-k}+\rho(x,y))^{1+\epsilon'}}+\frac{2^{-k\epsilon'}}{(2^{-k}+\rho(x,x_0))^{1+\epsilon'}}\Big]\\
&\qquad\times\frac{1}{(1+\rho(y,x_0))^{1+\gamma}}d\mu(y)\|f\|_{\mathcal{M}(\beta,\gamma)}\\
&\le C\Big[\frac{1}{(2^{-k}+\rho(x,x_0))^{1+\gamma}}+\frac{2^{-k\epsilon'}}{(2^{-k}+\rho(x,x_0))^{1+\epsilon'}}\frac{1}{(2^{-k}+\rho(x,x_0))^{\gamma}}\Big]\\
&\qquad\times\|f\|_{\mathcal{M}(\beta,\gamma)}\\
&\le C\frac{1}{(2^{-k}+\rho(x,x_0))^{1+\gamma}}\|f\|_{\mathcal{M}(\beta,\gamma)}\\
&\le C2^{\sigma_2 k}(1+\rho(x,x_0))^{-(1+\gamma')}\|f\|_{\mathcal{M}(\beta,\gamma)}
\end{aligned}
$$

where $k \le -M$ and $\sigma_2 = \gamma - \gamma' > 0$.

Putting these estimates on II_1 and II_2 into term II, we obtain

$$
\begin{aligned}
II &\le C\sum_{k<-M}2^{\sigma k}(1+\rho(x,x_0))^{-(1+\gamma')}\|f\|_{\mathcal{M}(\beta,\gamma)}\\
&\le C2^{-M\sigma}(1+\rho(x,x_0))^{-(1+\gamma')}\|f\|_{\mathcal{M}(\beta,\gamma)}
\end{aligned}
$$

where $\sigma > 0$.

These estimates imply (3.50). It remains to show (3.51). To do this, let $T = \sum_{|k|>M} D_k^N D_k$. Then it suffices to show that the operator T satisfies the conditions of Theorem 2.4. More precisely, using the same proof as in the proof of Lemma 3.17, one can show that the kernel of T satisfies the conditions in Theorem 2.4 with ϵ replaced by $\epsilon', 0 < \epsilon' < \epsilon$, and the constant C in these estimates is independent of M. Thus, applying Theorem 2.4 implies (3.51). We leave all details to the reader.

Finally, to see that the series in (3.29) converges in L^p for $1 < p < \infty$, by the proof above, we only need to show that (3.47) and (3.48) still hold with the norm of $\mathcal{M}(\beta', \gamma')$ replaced by the norm of L^p for $1 < p < \infty$. The estimates in Lemma 3.17 and the $T1$ theorem show that R_N is a Calderón-Zygmund operator with the operator norm at most $C2^{-N\delta}$ and, hence, R_N is bounded on L^p for $1 < p < \infty$ with the operator norm at most $C2^{-N\delta}$. This yields that (3.47) holds with the norm of $\mathcal{M}(\beta', \gamma')$ replaced by the norm of L^p for $1 < p < \infty$. To see that (3.48) still holds with the norm of $\mathcal{M}(\beta', \gamma')$ replaced by the norm of L^p for$1 < p < \infty$, it suffices to show that $\lim\limits_{M\to\infty} \| \sum\limits_{|k|>M} D_k^N D_k(f)\|_p = 0$ for $f \in L^p, 1 < p < \infty$. More precisely,

$$
\begin{aligned}
&\left\| \sum_{|k|>M} D_k^N D_k(f) \right\|_p \\
&\qquad = \sup_{\|g\|_{p'} \le 1} \left| \left\langle \sum_{|k|>M} D_k^N D_k(f), g \right\rangle \right| \\
&\qquad \le \sup_{\|g\|_{p'} \le 1} \left\| \left(\sum_{|k|>M} |D_k(f)|^2 \right)^{\frac{1}{2}} \right\|_p \cdot \left\| \left(\sum_{|k|>M} |(D_k^N)^*(g)|^2 \right)^{\frac{1}{2}} \right\|_{p'} \\
&\qquad \le C \sup_{\|g\|_{p'} \le 1} \left\| \left(\sum_{|k|>M} |D_k(f)|^2 \right)^{\frac{1}{2}} \right\|_p \|g\|_{p'} \\
&\qquad \le C \left\| \left(\sum_{|k|>M} |D_k(f)|^2 \right)^{\frac{1}{2}} \right\|_p
\end{aligned}
$$

where, by the Littlewood-Paley estimate on L^2 given in Section 1.3 and the Lebesgue dominated convergence theorem, the last term tends to zero as M tends to infinity. This ends the proof of Theorem 3.15.

We now consider the distribution spaces $(\dot{\mathcal{M}}_0(\beta, \gamma))'$, the dual of $\dot{\mathcal{M}}_0(\beta, \gamma)$, where $\dot{\mathcal{M}}_0(\beta, \gamma)$ is the closure of functions of $\mathcal{M}_0(\epsilon, \epsilon)$ in the norm of $\mathcal{M}(\beta, \gamma)$ with $0 < \beta, \gamma < \epsilon$, where ϵ is the regularity exponent in Definition 3.1. One should observe that $(\dot{\mathcal{M}}_0(\beta, \gamma))'$ is NOT a space of distributions but rather a space of distributions modulo constant functions. That will explain what is happening to the wavelet expansion of the function 1. By a duality argument, we obtain the following Calderón's identity.

Theorem 3.18. *Suppose that $\{D_k\}_{k\in\mathbb{Z}}, \{\widetilde{D}_k\}_{k\in\mathbb{Z}}$ and $\{\widetilde{\widetilde{D}}_k\}_{k\in\mathbb{Z}}$ are same as in Theorem 3.15. Then for $f \in (\dot{\mathcal{M}}_0(\beta, \gamma))'$,*

$$
f = \sum_k \widetilde{D}_k D_k(f) = \sum_k D_k \widetilde{\widetilde{D}}_k(f)
$$

where the series converge in the sense that for all $g \in \dot{\mathcal{M}}_0(\beta', \gamma')$ for $0 < \beta < \beta'$ and $0 < \gamma < \gamma'$,

$$\lim_{M\to\infty}\left\langle \sum_{|k|\le M} \widetilde{D}_k D_k(f), g\right\rangle = \langle f, g\rangle,$$

and

$$\lim_{M\to\infty}\left\langle \sum_{|k|\le M} D_k \widetilde{\widetilde{D}}_k(f), g\right\rangle = \langle f, g\rangle.$$

Theorem 3.15 do not provide expansions which converge in L^1 when f is integrable. If for instance the integral of f is 1, the convergence in L^1 would imply 1=0 because $\int D_k(x,y)d\mu(x) = \int \widetilde{D}_k(x,y)d\mu(x) = 0$.

Theorem 3.18 can be used to decompose arbitrary distributions $f \in (\dot{\mathcal{M}}_0(\beta,\gamma))'$, without worrying about their growth at infinity. But the expansion is only defined modulo constant functions. Once more if $f(x)$ is identically equal to 1, for all k, each $D_k(1)$ or $\widetilde{\widetilde{D}}_k(1)$ is zero because $\int D_k(x,y)d\mu(y) = \int \widetilde{\widetilde{D}}_k(x,y)d\mu(y) = 0$. This would give 1=0.

The difficulties we just described, however, disappear if a variant of Calderón's identity is being used. Instead of starting with the identity $I = \sum\limits_k D_k$, one instead has

$$I = S_{k_0} + \sum_{k>k_0} D_k \tag{3.54}$$

where $\{S_k\}_{k\in\mathbb{Z}}$ is an approximation to the identity, $D_k = S_k - S_{k-1}$ and k_0 is any fixed integer.

The following theorem gives this variant of Calderón's identities ([H4]).

Theorem 3.19. *Suppose that $\{S_k(x,y)\}_{k\in\mathbb{Z}}$ are father functions. Set $D_k(x,y) = S_k(x,y) - S_{k-1}(x,y)$ for all $k \ge 1, D_0(x,y) = S_0(x,y)$, and $D_k(x,y) = 0$ for all $k < 0$. Then there exist a positive integer N, and families of scaling functions $\{\widetilde{S}_k(x,y)\}$ and $\{\widetilde{\widetilde{S}}_k(x,y)\}$ for $0 \le k \le N$, and wavelets $\{\widetilde{D}_k(x,y)\}$, $\{\widetilde{\widetilde{D}}_k(x,y)\}$ for $k > N$ such that, for every $f \in \mathcal{M}(\beta,\gamma)$,*

$$\begin{aligned} f &= \sum_{0\le k\le N} \widetilde{S}_k D_k(f) + \sum_{k>N} \widetilde{D}_k D_k(f) \\ &= \sum_{0\le k\le N} D_k \widetilde{\widetilde{S}}_k(f) + \sum_{k>N} D_k \widetilde{\widetilde{D}}_k(f) \end{aligned} \tag{3.55}$$

where $\{\widetilde{S}_k\}$, $\{\widetilde{\widetilde{S}}_k\}$ and $\{\widetilde{D}_k\}$, $\{\widetilde{\widetilde{D}}_k\}$ are operators with kernels $\{\widetilde{S}_k(x,y)\}$, $\{\widetilde{\widetilde{S}}_k(x,y)\}$ and $\{\widetilde{D}_k(x,y)\}$, $\{\widetilde{\widetilde{D}}_k(x,y)\}$, respectively, the series converge in the norm of $\mathcal{M}(\beta',\gamma')$ for $0 < \beta' < \beta$ and $0 < \gamma' < \gamma$, and in the $L^p(X)$ spaces for $1 < p < \infty$. Moreover, $\widetilde{S}_k(x,y)$ and $\widetilde{D}_k(x,y)$ satisfy the conditions (3.14) and (3.15) with ϵ replaced by $\epsilon', 0 < \epsilon' < \epsilon$, and $\int \widetilde{S}_k(x,y)d\mu(y) = \int \widetilde{S}_k(x,y)d\mu(x) = 1$ for $0 \le k \le N, \int \widetilde{D}_k(x,y)d\mu(y) =$

$\int \widetilde{D}_k(x,y)d\mu(x) = 0$ for all $k > N$, and $\widetilde{\widetilde{S}}_k(x,y)$ and $\widetilde{\widetilde{D}}_k(x,y)$ satisfy the conditions (3.14) and (3.16) with ϵ replaced by $\epsilon', 0 < \epsilon' < \epsilon$, and $\int \widetilde{\widetilde{S}}_k(x,y)d\mu(y) = \int \widetilde{\widetilde{S}}_k(x,y)d\mu(x) = 1$ for $0 \le k \le N$, $\int \widetilde{\widetilde{D}}_k(x,y)d\mu(y) = \int \widetilde{\widetilde{D}}_k(x,y)d\mu(x) = 0$ for all $k > N$.

By a duality argument, the series in (3.55) also converge in $(\dot{\mathcal{M}}(\beta',\gamma'))', \beta < \beta', \gamma < \gamma'$, where $(\dot{\mathcal{M}}(\beta',\gamma'))'$ is the dual of $(\dot{\mathcal{M}}(\beta',\gamma')$, the closure of functions of $\mathcal{M}(\epsilon,\epsilon)$ in the norm of $\mathcal{M}(\beta',\gamma')$ with $0 < \beta',\gamma' < \epsilon$.

We remark that the series in (3.55) cannot converge for the L^1 norm by the same reason as given above. But, (3.55) still holds for $f(x)$ being identically equal to 1 since $D_0(1) = \widetilde{S}_k(1) = \widetilde{\widetilde{S}}_k(1) = 1$ for $0 \le k \le N$, and $D_k(1) = 0$ for all $k \ge 1$ and $\widetilde{\widetilde{D}}_k(1) = 0$ for all $k > N$.

To show Theorem 3.19, we need the following lemmas, which are similar to Proposition 3.16 and Lemma 3.17.

Proposition 3.20 *Suppose that $\{S_k\}_{k\in\mathbb{Z}}$ is an approximation to the identity and its kernels satisfy the condition (3.19) with $\sigma = \epsilon \le \theta$. Set $D_k = S_k - S_{k-1}$ for all $k \ge 1, D_0 = S_0$, and $D_k = 0$ for all $k < 0$. Let $T_N = \sum\limits_{k\ge 0} D_k^N D_k$ where $D_k^N = \sum\limits_{|j|\le N} D_{k+j}$ and N is a fixed positive integer. Then T_N^{-1} exists if N is a sufficiently large integer. Moreover, there exists a constant C such that for $f \in \mathcal{M}_0(x_1,r,\beta,\gamma)$ with $x_1 \in X$, $r > 0$ and $0 < \beta,\gamma < \epsilon$,*

$$\|T_N^{-1}(f)\|_{\mathcal{M}(x_1,r,\beta,\gamma)} \le C\|f\|_{\mathcal{M}(x_1,r,\beta,\gamma)}.$$

The proof of Proposition 3.20 is based on the following lemma.

Lemma 3.21. *Suppose that $\{S_k\}_{k\in\mathbb{Z}}$ is an approximation to the identity and its kernels satisfy the condition (3.19) with $\sigma = \epsilon \le \theta$. Set $D_k = S_k - S_{k-1}$ for all $k \ge 1, D_0 = S_0$, and $D_k = 0$ for all $k < 0$. Let $R_N = \sum\limits_{|l|>N}\sum\limits_{k\ge 0} D_{k+l}D_k$. Then for $0 < \epsilon' < \epsilon$, $R_N \in CZK(\epsilon') \cap SWBP$, $R_N(1) = (R_N)^*(1) = 0$. Moreover, $R_N(x,y)$, the kernels of R_N satisfy the following estimates: there exist a constants $0 < C < \infty$ which is independent of N, and $\delta > 0$ such that*

$$|R_N(x,y)| \le C2^{-N\delta}\rho(x,y)^{-1};$$

$$|R_N(x,y) - R_N(x',y)| \le C2^{-N\delta}\rho(x,x')^{\epsilon'}\rho(x,y)^{-(1+\epsilon')}$$

for $\rho(x,x') \le \frac{1}{2A}\rho(x,y)$;

$$|R_N(x,y) - R_N(x,y')| \le C2^{-N\delta}\rho(y,y')^{\epsilon'}\rho(x,y)^{-(1+\epsilon')}$$

for $\rho(y,y') \le \frac{1}{2A}\rho(x,y)$;

$$|R_N(x,y) - R_N(x',y) - R_N(x,y') + R_N(x',y')|$$

$$\leq C2^{-N\delta}\rho(x,x')^{\epsilon'}\rho(y,y')^{\epsilon'}\rho(x,y)^{-(1+2\epsilon')}$$

for $\rho(x,x') \leq \frac{1}{2A}\rho(x,y)$ *and* $\rho(y,y') \leq \frac{1}{2A}\rho(x,y)$;

$$|\langle R_N, f\rangle| \leq C2^{-N\delta}r$$

for all $f \in C_0^\eta(X \times X)$ *with* $\text{supp}(f) \subseteq B(x_1,r) \times B(y_1,r), x_1$ *and* $y_1 \in X, \|f\|_\infty \leq 1, \|f(\cdot,y)\|_\eta \leq r^{-\eta}$, *and* $\|f(x,\cdot)\|_\eta \leq r^{-\eta}$ *for all* x *and* $y \in X$.

It is clear that Proposition 3.20 follows immediately from Lemma 3.21. To show Lemma 3.21, note that there are only three cases in the definition of R_N : (i) $D_{k+l}D_k$ with $k+l \geq 1$ and $k \geq 1$; (ii) $D_{k+l}D_k = S_0D_k$ with $k+l=0$ and $k \geq 1$; (iii) $D_{k+l}D_k = D_lS_0$ with $l \geq 1$ and $k=0$. So we can handle case (i) as in Lemma 3.17. To deal with cases (ii) and (iii), by the remarks in Section 3.2, we can obtain the same estimates as in the case (i), and, hence, this implies Lemma 3.21. We leave all details to the reader.

We now return to the proof of Theorem 3.19. Since the proofs of the two identities in Theorem 3.19 are similar, so we only show the first identity in (3.55). Fix a large integer N such that Proposition 3.20 holds. It is easy to check that $D_k(\cdot,y)$, the kernel of D_k, is in $\mathcal{M}_0(y,2^{-k},\epsilon,\epsilon)$ for all $k \geq 1$ and $D_k^N(\cdot,y)$, the kernel of D_k^N, is in $\mathcal{M}_0(y,2^{-k},\epsilon,\epsilon)$ for all $k > N$. Set $\widetilde{S}_k = T_N^{-1}(D_k^N)$ for $0 \leq k \leq N$, and $\widetilde{D}_k = T_N^{-1}(D_k^N)$ for $k > N$, where T_N^{-1} is defined as in Proposition 3.20. Note that $D_k^N = \sum_{|j|\leq N} D_{k+j} = \sum_{0\leq j\leq k+N} D_j$ for $0 \leq k \leq N$ and $D_k^N = \sum_{|j|\leq N} D_{k+j} = \sum_{k-N\leq j\leq k+N} D_j$ for $k > N$. By Proposition 3.20, $\widetilde{D}_k \in \mathcal{M}_0(y,2^{-k},\epsilon',\epsilon')$ with $0< \epsilon' < \epsilon$ and this implies that $\widetilde{D}_k(x,y)$, the kernels of $\widetilde{D}_k$, satisfy conditions (3.14) and (3.15) with ϵ replaced by ϵ', and $\int \widetilde{D}_k(x,y)d\mu(x) = 0$. The fact that $(T_N^{-1})^*(1) = 1$ yields $\int \widetilde{D}_k(x,y)d\mu(y) = \int D_k^N(x,y)d\mu(y) = 0$ for $k > N$.

We know that

$$\begin{aligned}\widetilde{S}_k = T_N^{-1}(D_k^N) = T_N^{-1}\Bigg(\sum_{0\leq j\leq k+N} D_j\Bigg) &= \sum_{0\leq j\leq k+N} T_N^{-1}(D_j)\\ = T_N^{-1}(S_0) + \sum_{1\leq j\leq k+N} T_N^{-1}(D_j)\end{aligned}$$

for $0 \leq k \leq N$. In order to prove that $\widetilde{S}_k(x,y)$, the kernels of $\widetilde{S}_k$, satisfy the conditions (3.14) and (3.15) with ϵ replaced by ϵ' and $\int \widetilde{S}_k(x,y)d\mu(y) = \int \widetilde{S}_k(x,y)d\mu(x) = 1$, it suffices to prove that $T_N^{-1}S_0(x,y)$, the kernel of $T_N^{-1}S_0$, satisfies the conditions (3.14) and (3.15) with ϵ replaced by ϵ' and $\int T_N^{-1}S_0(x,y)d\mu(y) = \int T_N^{-1}S_0(x,y)d\mu(x) = 1$ since, by Proposition 3.20, $T_N^{-1}(D_j) \in \mathcal{M}_0(y,2^{-j},\epsilon',\epsilon')$ for $0 < \epsilon' < \epsilon$ and $j > 0$. To estimate $T_N^{-1}S_0(x,y)$, Theorem 2.4 cannot be applied because $S_0(x,y)$ is not in $\mathcal{M}_0(y,1,\epsilon',\epsilon')$. However, we claim that $(R_N)^nS_0(x,y)$, the kernels of

$(R_N)^n S_0$, satisfy the following estimates: For $0 < \epsilon' < \epsilon$ and $n \geq 1$, there exist constants C and $\delta > 0$ such that

$$|(R_N)^n S_0(x,y)| \leq (C2^{-N\delta})^n \frac{1}{(1+\rho(x,y))^{1+\epsilon}}, \tag{3.56}$$

$$\begin{aligned} &|(R_N)^n S_0(x,y) - (R_N)^n S_0(x',y)| \\ &\quad \leq (C2^{-N\delta})^n \Big(\frac{\rho(x,x')}{1+\rho(x,y)}\Big)^{\epsilon'} \frac{1}{(1+\rho(x,x_0))^{1+\epsilon'}} \end{aligned} \tag{3.57}$$

for $\rho(x,x') \leq \frac{1}{2A}(1+\rho(x,y)$.

Inequalities (3.56) and (3.57), together with the following fact

$$T_N^{-1} S_0(x,y) = \sum_{n=0}^{\infty} (R_N)^n S_0(x,y),$$

imply that if N is sufficiently large, then $T_N^{-1}S_0(x,y)$ satisfies the conditions (3.14) and (3.15) with ϵ replaced by ϵ'. Then $\int T_N^{-1} S_0(x,y) d\mu(x) = \int T_N^{-1} S_0(x,y) d\mu(y) = 1$ follows from the facts that $\int S_0(x,y) d\mu(x) = \int S_0(x,y) d\mu(y) = 1$ and $R_N(1) = (R_N)^*(1) = 0$.

We now prove (3.56) and (3.57). To do this, we first write $(R_N)^n S_0(x,y)$, the kernel of $(R_N)^n S_0$, as follows:

$$\begin{aligned} (R_N)^n S_0(x,y) = &\sum_{|j_n|>N} \sum_{k_n \geq 0} \cdots \\ &\times \sum_{|j_2|>N} \sum_{k_2\geq 0} \sum_{|j_1|>N} \sum_{k_1 \geq 0} D_{k_n+j_n} D_{k_n} \cdots D_{k_2+j_2} D_{k_2} D_{k_1+j_1} D_{k_1} S_0(x,y). \end{aligned}$$

So it suffices to show the following estimates:

$$\begin{aligned} &|D_{k_n+j_n} D_{k_n} \cdots D_{k_2+j_2} D_{k_2} D_{k_1+j_1} D_{k_1} S_0(x,y)| \\ &\quad \leq C^n 2^{-k_n\epsilon'} 2^{-k_{n-1}\epsilon'} \cdots 2^{-k_1\epsilon'} \frac{1}{(1+\rho(x,y))^{1+\epsilon}}, \end{aligned} \tag{3.58}$$

$$\begin{aligned} &|D_{k_n+j_n} D_{k_n} \cdots D_{k_2+j_2} D_{k_2} D_{k_1+j_1} D_{k_1} S_0(x,y)| \\ &\quad \leq C^n 2^{-|j_n|\epsilon'} 2^{-|j_{n-1}|\epsilon'} \cdots 2^{-|j_1|\epsilon'} \frac{1}{(1+\rho(x,y))^{1+\epsilon}}, \end{aligned} \tag{3.59}$$

$$\begin{aligned} &|D_{k_n+j_n} D_{k_n} \cdots D_{k_2+j_2} D_{k_2} D_{k_1+j_1} D_{k_1} S_0(x,y) \\ &\quad - D_{k_n+j_n} D_{k_n} \cdots D_{k_2+j_2} D_{k_2} D_{k_1+j_1} D_{k_1} S_0(x',y)| \\ &\leq C^n \Big(\frac{\rho(x,x')}{1+\rho(x,y)}\Big)^{\epsilon} \frac{1}{(1+\rho(x,x_0))^{1+\epsilon}} \end{aligned} \tag{3.60}$$

for $\rho(x,x') \leq \frac{1}{4A^2}(1+\rho(x,y)$.

Assuming these estimates for the moment, from (3.58) and (3.59), we obtain

$$|D_{k_n+j_n}D_{k_n}\cdots D_{k_2+j_2}D_{k_2}D_{k_1+j_1}D_{k_1}S_0(x,y)| \tag{3.61}$$
$$\leq C^n 2^{-\frac{1}{2}|j_n|\epsilon'}2^{-\frac{1}{2}|j_{n-1}|\epsilon'}\ldots 2^{-\frac{1}{2}|j_1|\epsilon'}2^{-\frac{1}{2}k_n\epsilon'}2^{-\frac{1}{2}k_{n-1}\epsilon'}\ldots 2^{-\frac{1}{2}k_1\epsilon'}$$
$$\times\frac{1}{(1+\rho(x,y))^{1+\epsilon}},$$

$$|D_{k_n+j_n}D_{k_n}\cdots D_{k_2+j_2}D_{k_2}D_{k_1+j_1}D_{k_1}S_0(x,y) \tag{3.62}$$
$$-D_{k_n+j_n}D_{k_n}\cdots D_{k_2+j_2}D_{k_2}D_{k_1+j_1}D_{k_1}S_0(x',y)|$$
$$\leq C^n 2^{-\frac{1}{2}|j_n|\epsilon'}2^{-\frac{1}{2}|j_{n-1}|\epsilon'}\ldots 2^{-\frac{1}{2}|j_1|\epsilon'}2^{-\frac{1}{2}k_n\epsilon'}2^{-\frac{1}{2}k_{n-1}\epsilon'}\ldots 2^{-\frac{1}{2}k_1\epsilon'}$$
$$\times\frac{1}{(1+\rho(x,y))^{1+\epsilon}}$$

for $\rho(x,x') \leq \frac{1}{4A^2}(1+\rho(x,y)$.

Taking the geometric mean of (3.60) and (3.62) yields

$$|D_{k_n+j_n}D_{k_n}\cdots D_{k_2+j_2}D_{k_2}D_{k_1+j_1}D_{k_1}S_0(x,y) \tag{3.63}$$
$$-D_{k_n+j_n}D_{k_n}\cdots D_{k_2+j_2}D_{k_2}D_{k_1+j_1}D_{k_1}S_0(x',y)|$$
$$\leq C^n 2^{-|j_n|\delta}2^{-|j_{n-1}|\delta}\ldots 2^{-|j_1|\delta}2^{-k_n\delta}2^{-k_{n-1}\delta}\ldots 2^{-k_1\delta}\Big(\frac{\rho(x,x')}{1+\rho(x,y)}\Big)^{\epsilon'}$$
$$\times\frac{1}{(1+\rho(x,y))^{1+\epsilon'}}$$

for $\rho(x,x') \leq \frac{1}{4A^2}(1+\rho(x,y), 0<\epsilon'<\epsilon$, and $\delta>0$.

Summing over $k_1, k_2, \cdots, k_n$, and then $j_1, j_2, \cdots, j_n$, (3.61) implies (3.56) and (3.63) implies (3.57), respectively. So it remains to show (3.58), (3.59) and (3.60). Consider $n=1$ first. Following the proof of Proposition 3.16 and Lemma 3.9, $D_{k_1}S_0(x,y)$, the kernel of $D_{k_1}S_0$, satisfies the following estimates: For $k_1 \geq 0$

$$|D_{k_1}S_0(x,y)| \leq C2^{-k_1\epsilon'}\frac{1}{(1+\rho(x,y))^{1+\epsilon}},$$

$$|D_{k_1}S_0(x,y) - D_{k_1}S_0(x',y)| \leq C\Big(\frac{\rho(x,x')}{1+\rho(x,y)}\Big)^{\epsilon}\frac{1}{(1+\rho(x,y))^{1+\epsilon}}$$

for $\rho(x,x') \leq \frac{1}{4A^2}(1+\rho(x,y)$.

The estimates above imply, again, that for $k_1 \geq 0$ and $k_1+j_1 \geq 0$

$$|D_{k_1+j_1}D_{k_1}S_0(x,y)| \leq C2^{-(k_1+j_1)\epsilon'}2^{-k_1\epsilon'}\frac{1}{(1+\rho(x,y))^{1+\epsilon}}$$
$$\leq C2^{-k_1\epsilon'}\frac{1}{(1+\rho(x,y))^{1+\epsilon}},$$

$$|D_{k_1+j_1}D_{k_1}S_0(x,y)-D_{k_1+j_1}D_{k_1}S_0(x',y)| \le C\Big(\frac{\rho(x,x')}{1+\rho(x,y)}\Big)^{\epsilon}\frac{1}{(1+\rho(x,y))^{1+\epsilon}}$$

for $\rho(x,x') \le \frac{1}{4A^2}(1+\rho(x,y)$.

Similarly, estimating first the kernel of $D_{k_1+j_1}D_{k_1}$ and then the kernel of $D_{k_1+j_1}D_{k_1}S_0$, we obtain

$$\begin{aligned}|D_{k_1+j_1}D_{k_1}S_0(x,y)| &\le C2^{-[(k_1+j_1)\wedge k_1]\epsilon'}2^{-|j_1|\epsilon'}\frac{1}{(1+\rho(x,y))^{1+\epsilon}}\\ &\le C2^{-|j_1|\epsilon'}\frac{1}{(1+\rho(x,y))^{1+\epsilon}}\end{aligned}$$

for $k_1 \ge 0$ and $k_1+j_1 \ge 0$. These estimates imply (3.58), (3.59) and (3.60) with $n=1$. Repeating the above proof we can show that the estimates (3.58), (3.59) and (3.60) hold for all $n \ge 1$. The proof used for verifying the conditions for $\widetilde{\widetilde{S}}_k$ and $\widetilde{\widetilde{D}}_k$ is similar. We leave these details to the reader.

We now prove that the series in (3.55) converges in the norm of $\mathcal{M}(\beta',\gamma')$ for $0<\beta'<\beta$ and $0<\gamma'<\gamma$. To do this, suppose $f\in\mathcal{M}(\beta,\gamma)$. Then

$$\begin{aligned}&\sum_{0\le k\le N}\widetilde{S}_kD_k(f)+\sum_{N+1\le k\le M}\widetilde{D}_kD_k(f)\\ &\quad= T_N^{-1}\Big(\sum_{0\le k\le M}D_k^ND_k\Big)(f)\\ &\quad= T_N^{-1}\Big(T_N-\sum_{k>M}D_k^ND_k\Big)(f)\\ &\quad= T_N^{-1}T_N(f)-T_N^{-1}\Big(\sum_{k>M}D_k^ND_k\Big)(f)\\ &\quad= f-\lim_{j\to\infty}(R_N)^j(f)-T_N^{-1}\Big(\sum_{k>M}D_k^ND_k\Big)(f).\end{aligned}$$

Thus,

$$\begin{aligned}&\Big\|\sum_{0\le k\le N}\widetilde{S}_kD_k(f)+\sum_{N+1\le k\le M}\widetilde{D}_kD_k(f)-f\Big\|_{\mathcal{M}(\beta',\gamma')}\\ &\quad\le \lim_{j\to\infty}\|(R_N)^j(f)\|_{\mathcal{M}(\beta',\gamma')}+\Big\|T_N^{-1}\Big(\sum_{k>M}D_k^ND_k\Big)(f)\Big\|_{\mathcal{M}(\beta',\gamma')}.\end{aligned}$$

By a similar proof as given in (3.47) and the fact that $(R_N)^*(1)=0$, it is easy to see that $R_N(f)(x)\in\mathcal{M}(\beta'',\gamma'')$ with $\beta'<\beta''<\beta,\gamma'<\gamma''<\gamma$.

By Lemma 3.17 and Theorem 2.4, we obtain

$$\|(R_N)^j(f)\|_{\mathcal{M}(\beta',\gamma')} \le (C2^{-N\delta})^j \|f\|_{\mathcal{M}(\beta',\gamma')} \tag{3.64}$$

which implies $\lim_{j\to\infty} \|(R_N)^j(f)\|_{\mathcal{M}(\beta',\gamma')} = 0$.

To prove that $\left\|T_N^{-1}(\sum_{k>M} D_k^N D_k)(f)\right\|_{\mathcal{M}(\beta',\gamma')}$ tends to zero as M tends to infinity, it suffices to show the following estimate:

$$\left\|\left(\sum_{k>M} D_k^N D_k\right)(f)\right\|_{\mathcal{M}(\beta',\gamma)} \le C2^{-\sigma M}\|f\|_{\mathcal{M}(\beta,\gamma)} \tag{3.65}$$

for all $0 < \beta' < \beta$ and some $\sigma > 0$, and a constant C which is independent of f and M.

In fact, we will show that for $0 < \beta' < \beta'' < \beta$ and some $\sigma > 0$, there exists a constant C which is independent of f and M such that

$$\left|\sum_{k>M} D_k^N D_k(f)(x)\right| \le C2^{-\beta M}(1+\rho(x,x_0))^{-(1+\gamma)}\|f\|_{\mathcal{M}(\beta,\gamma)}, \tag{3.66}$$

$$\begin{aligned}\left|\sum_{k>M} D_k^N D_k(f)(x) - \sum_{k>M} D_k^N D_k(f)(x')\right| & \\ \le C\Big(\frac{\rho(x,x')}{1+\rho(x,x_0)}\Big)^{\beta''} (1+\rho(x,x_0))^{-(1+\gamma)}\|f\|_{\mathcal{M}(\beta,\gamma)}\end{aligned} \tag{3.67}$$

for $\rho(x,x') \le \frac{1}{4A^2}(1+\rho(x,x_0)$.

To see this, note that if $\rho(x,x') \le \frac{1}{4A^2}(1+\rho(x,x_0))$ then, by (3.66),

$$\begin{aligned}\left|\sum_{k>M} D_k^N D_k(f)(x) - \sum_{k>M} D_k^N D_k(f)(x')\right| & \\ \le C2^{-\beta M}(1+\rho(x,x_0))^{-(1+\gamma)}\|f\|_{\mathcal{M}(\beta,\gamma)}.\end{aligned} \tag{3.68}$$

Taking the geometric mean of (3.67) and (3.68), we obtain

$$\begin{aligned}\left|\sum_{k>M} D_k^N D_k(f)(x) - \sum_{k>M} D_k^N D_k(f)(x')\right| & \\ \le C2^{-\sigma M}\Big(\frac{\rho(x,x')}{1+\rho(x,x_0)}\Big)^{\beta'} (1+\rho(x,x_0))^{-(1+\gamma)}\|f\|_{\mathcal{M}(\beta,\gamma)}\end{aligned} \tag{3.69}$$

for $\rho(x,x') \le \frac{1}{4A^2}(1+\rho(x,x_0)$, which together with (3.66) implies (3.65).

We first prove (3.66). Denote $E_k = D_k^N D_k$. It is easy to check that $E_k(x,y)$, the kernel of E_k, satisfies the condition (3.14), (3.15) and (3.16) with ϵ replaced by $\epsilon', 0 < \beta, \gamma < \epsilon' < \epsilon$, and $E_k(1) = 0$ for $k \geq 1$. Then

$$
\begin{aligned}
\Big|\sum_{k>M} D_k^N D_k(f)(x)\Big| &= \Big|\sum_{k>M} E_k(f)(x)\Big| \\
&= \Big|\sum_{k>M} \int E_k(x,y)[f(y)-f(x)]d\mu(y)\Big| \\
&\leq \sum_{k>M} \int_{\rho(x,y)\leq \frac{1}{2A}(1+\rho(x,x_0))} |E_k(x,y)||f(y)-f(x)|d\mu(y) \\
&\quad + \sum_{k>M} \int_{\rho(x,y)> \frac{1}{2A}(1+\rho(x,x_0))} |E_k(x,y)||f(y)-f(x)|d\mu(y) \\
&\leq C\sum_{k>M} \int_{\rho(x,y)\leq \frac{1}{2A}(1+\rho(x,x_0))} |E_k(x,y)|\Big(\frac{\rho(x,y)}{1+\rho(x,x_0)}\Big)^{\beta} \\
&\quad \times (1+\rho(x,x_0))^{-(1+\gamma)}\|f\|_{\mathcal{M}(\beta,\gamma)} d\mu(y) \\
&\quad + C\sum_{k>M} \int_{\rho(x,y)> \frac{1}{2A}(1+\rho(x,x_0))} |E_k(x,y)|[(1+\rho(x,y))^{-(1+\gamma)} \\
&\qquad + (1+\rho(x,x_0))^{-(1+\gamma)}]\|f\|_{\mathcal{M}(\beta,\gamma)} d\mu(y) \\
&\leq C\sum_{k>M} \{2^{-\beta k}(1+\rho(x,x_0))^{-(1+\gamma)} + 2^{-k\epsilon'}(1+\rho(x,x_0))^{-(1+\epsilon')} \\
&\quad + 2^{-k\epsilon'}(1+\rho(x,x_0))^{-(1+\gamma)}\}\|f\|_{\mathcal{M}(\beta,\gamma)} \\
&\leq C2^{-\beta M}(1+\rho(x,x_0))^{-(1+\gamma)}\|f\|_{\mathcal{M}(\beta,\gamma)},
\end{aligned}
$$

from which (3.66) follows.

Now we prove (3.67). Consider the case where $\rho(x,x') \leq \frac{1}{4A^2}(1+\rho(x,x_0)$. We write

$$
\begin{aligned}
\Big|\sum_{k>M} D_k^N D_k(f)(x) - \sum_{k>M} D_k^N D_k(f)(x')\Big| & \\
&= \Big|\sum_{k>M} \int [E_k(x,y)-E_k(x',y)]f(y)d\mu(y)\Big| \\
&= \Big|\sum_{k>M} \int [E_k(x,y)-E_k(x',y)][f(y)-f(x)]d\mu(y)\Big| \\
&\leq \sum_{k>M} \int_{W_1} |E_k(x,y)-E_k(x',y)||f(y)-f(x)|d\mu(y) \\
&\quad + \sum_{k>M} \int_{W_2} |E_k(x,y)-E_k(x',y)||f(y)-f(x)|d\mu(y)
\end{aligned}
$$

$$+\sum_{k>M}\int_{W_3}|E_k(x,y)-E_k(x',y)||f(y)-f(x)|d\mu(y)$$
$$:= I+II+III$$

where

$$W_1=\Big\{z\in X:\rho(x,x')\le\frac{1}{4A^2}(1+\rho(x,x_0))\le\frac{1}{2A}(2^{-k}+\rho(x,y))\Big\},$$
$$W_2=\Big\{z\in X:\rho(x,x')\le\frac{1}{2A}(2^{-k}+\rho(x,y))\le\frac{1}{4A^2}(1+\rho(x,x_0))\Big\},$$
$$W_3=\Big\{z\in X:\rho(x,x')>\frac{1}{2A}(2^{-k}+\rho(x,y))\Big\}.$$

By the smoothness conditions on E_k and f, we obtain

$$II\le C\sum_{k>M}\int_{W_2}\Big(\frac{\rho(x,x')}{2^{-k}+\rho(x,y)}\Big)^{\epsilon'}\frac{2^{-k\epsilon'}}{(2^{-k}+\rho(x,y))^{1+\epsilon'}}\Big(\frac{\rho(x,y)}{1+\rho(x,x_0)}\Big)^{\beta}$$
$$\times(1+\rho(x,x_0))^{-(1+\gamma)}\|f\|_{\mathcal{M}(\beta,\gamma)}d\mu(y)$$
$$\le C\sum_{k>M}\Big(\frac{\rho(x,x')}{1+\rho(x,x_0)}\Big)^{\beta'}(1+\rho(x,x_0))^{-(1+\gamma)}\|f\|_{\mathcal{M}(\beta,\gamma)}$$
$$\times\int_{W_2}\frac{2^{-k\epsilon'}}{(2^{-k}+\rho(x,y))^{1+\epsilon'}}(\rho(x,y))^{(\beta-\beta')}d\mu(y)$$
$$\le C\sum_{k>M}2^{-k(\beta-\beta')}\Big(\frac{\rho(x,x')}{1+\rho(x,x_0)}\Big)^{\beta'}(1+\rho(x,x_0))^{-(1+\gamma)}\|f\|_{\mathcal{M}(\beta,\gamma)}$$
$$\le C\Big(\frac{\rho(x,x')}{1+\rho(x,x_0)}\Big)^{\beta'}(1+\rho(x,x_0))^{-(1+\gamma)}\|f\|_{\mathcal{M}(\beta,\gamma)}.$$

Using the smoothness condition on E_k and size condition on f,

$$I\le C\sum_{k>M}\int_{W_1}\Big(\frac{\rho(x,x')}{2^{-k}+\rho(x,y)}\Big)^{\epsilon'}\frac{2^{-k\epsilon'}}{(2^{-k}+\rho(x,y))^{1+\epsilon'}}\{|f(y)|+|f(x)|\}d\mu(y)$$
$$\le C\sum_{k>M}\Big(\frac{\rho(x,x')}{1+\rho(x,x_0)}\Big)^{\epsilon'}\frac{2^{-k\epsilon'}}{(1+\rho(x,x_0))^{1+\epsilon'}}\int_{W_1}|f(y)|d\mu(y)$$
$$+C\sum_{k>M}\Big(\frac{\rho(x,x')}{1+\rho(x,x_0)}\Big)^{\epsilon'}|f(x)|\int_{W_1}\frac{2^{-k\epsilon'}}{(2^{-k}+\rho(x,y))^{1+\epsilon'}}d\mu(y)$$
$$\le C\Big(\frac{\rho(x,x')}{1+\rho(x,x_0)}\Big)^{\beta'}(1+\rho(x,x_0))^{-(1+\gamma)}\|f\|_{\mathcal{M}(\beta,\gamma)}.$$

The size condition on E_k and smoothness on f yield

$$
\begin{aligned}
III &\le C\sum_{k>M}\int_{W_3}\{|E_k(x,y)|+|E_k(x'.y)|\}\Big(\frac{\rho(x,y)}{1+\rho(x,x_0)}\Big)^{\beta}\\
&\qquad\times(1+\rho(x,x_0))^{-(1+\gamma)}\|f\|_{\mathcal{M}(\beta,\gamma)}d\mu(y)\\
&\le C\sum_{k>M}\Big(\frac{\rho(x,x')}{1+\rho(x,x_0)}\Big)^{\beta'}(1+\rho(x,x_0))^{-(1+\gamma)}\|f\|_{\mathcal{M}(\beta,\gamma)}\\
&\qquad\times\int_{W_3}\{|E_k(x,y)|+|E_k(x',y)|\}\rho(x,y)^{(\beta-\beta')}d\mu(y)\\
&\le C\Big(\frac{\rho(x,x')}{1+\rho(x,x_0)}\Big)^{\beta'}(1+\rho(x,x_0))^{-(1+\gamma)}\|f\|_{\mathcal{M}(\beta,\gamma)}
\end{aligned}
$$

which implies (3.67).

Finally, to see that the series in (3.55) converges in L^p for $1<p<\infty$ we use the above proof and we only need to show that the last two terms in (3.64) tend to zero as M tends to infinity when the norm of $\mathcal{M}(\beta',\gamma')$ is replaced by the norm of L^p for $1<p<\infty$. The estimates in Lemma 3.17, as we have proved above, imply that R_N is a Calderón-Zygmund operator with the operator norm at most $C2^{-N\delta}$ and, hence, R_N is bounded on L^p for $1<p<\infty$ with the operator norm at most $C2^{-N\delta}$. This also implies that T_N^{-1} is bounded on L^p for $1<p<\infty$. So it suffices to show $\lim_{M\to\infty}\|\sum_{k>M}D_k^N D_k(f)\|_p=0$ for $f\in L^p, 1<p<\infty$. More precisely,

$$
\begin{aligned}
&\Big\|\sum_{k>M}D_k^N D_k(f)\Big\|_p\\
&\quad=\sup_{\|g\|_{p'}\le 1}\Big|\Big\langle\sum_{k>M}D_k^N D_k(f),g\Big\rangle\Big|\\
&\quad=\le\sup_{\|g\|_{p'}\le 1}\Big\|\Big(\sum_{k>M}|D_k(f)|^2\Big)^{\frac12}\Big\|_p\cdot\Big\|\Big(\sum_{k>M}|(D_k^N)^*(g)|^2\Big)^{\frac12}\Big\|_{p'}\\
&\quad=\le C\sup_{\|g\|_{p'}\le 1}\Big\|\Big(\sum_{k>M}|D_k(f)|^2\Big)^{\frac12}\Big\|_p\|g\|_{p'}\le C\Big\|\Big(\sum_{k>M}|D_k(f)|^2\Big)^{\frac12}\Big\|_p
\end{aligned}
$$

where by the Littlewood-Paley estimates on $L^p, 1<p<\infty$ the last term tends to zero as M tends to infinity. This ends the proof of Theorem 3.19.

In the next section a discrete version of Calderón's identity will provide wavelet expansions on spaces of homogeneous type.

3.5 Wavelet Expansions on Spaces of Homogeneous Type

In this section, we provide wavelet expansions on $L^2(X)$ and prove that these wavelet expansions still hold on $L^p(X), 1 < p < \infty$, and distribution spaces.

Let us first return to the standard case of $\mathbb{R}^n$. As we mentioned in Section 3.4, let $\psi \in L^1(\mathbb{R}^n)$ be a function whose integral is zero and whose Fourier transform $\widehat{\psi}(\xi)$ satisfies

$$\int_0^\infty |\widehat{\psi}(t\xi)|^2 \frac{dt}{t} = 1$$

for each $\xi \neq 0$. We use ψ to define wavelets $\psi_{(a,b)}, a > 0, b \in \mathbb{R}^n$, by $\psi_{(a,b)}(x) = a^{-\frac{n}{2}}\psi(\frac{x-b}{a})$. Notice that this definition is inconsistent with the one used in (3.30). We define the "wavelet transformation" of $f \in L^2(\mathbb{R}^n)$ by $F(a,b) = \langle f, \psi_{(a,b)} \rangle$, as a function of a and b. We obtain a continuous version of wavelet expansion on $L^2(\mathbb{R}^n)$:

$$f(x) = \int_0^\infty \int_{\mathbb{R}^n} F(a,b)\psi_{(a,b)} db \frac{da}{a^{1+n}}.$$

Finally we can replace the above identity by a discrete version, obtained by paving $\mathbb{R}^n$ by "Whitney cubes". We then summarily replace a by 2^{-j} and b by $k2^{-j}, k \in \mathbb{Z}^n$, and $dadb$ by the volume of the Whitney cube, that is, $2^{-(n+1)j}$. This volume exactly compensates for a^{1+n}, and the above identity becomes

$$f(x) \approx \sum_j \sum_k \alpha(j,k) 2^{\frac{nj}{2}} \psi(2^j x - k).$$

One should observe that using the theory of frames, one can give an exact formula on iteration which has been developed by Daubechies ([Da]) as we described in Section 3.2. But, however, this iteration method, in general, cannot provide an identity with a wavelet expansion. The method we will use is the theory of Calderón-Zygmund operators as used in the previous section for proving Calderón's identity.

We now come to a general space of homogeneous type. One needs an analogue of the Whitney cubes on spaces of homogeneous type. This is the following result given by M. Christ [Ch2], which provides an analogue of the grid of Euclidean dyadic cubes on spaces of homogeneous type.

Theorem 3.22. *For every integer $k \in \mathbb{Z}$, there exist a collection of open subsets $\{Q_\tau^k \subseteq X : k \in \mathbb{Z}, \tau \in I_k\}$, where I_k denotes some (possibly finite) index set depending on k, and a constant $\delta \in (0,1)$, $a_0 > 0$, $\eta > 0$ and $0 < c_1$, $c_2 < \infty$ such that*

$$\mu(\{X \setminus \cup Q_\tau^k\}) = 0$$

for all $k \in \mathbb{Z}$.

If $l \geq k$ then either

$$Q^l_{\tau'} \subseteq Q^k_\tau$$

or

$$Q^l_{\tau'} \cap Q^k_\tau = \phi.$$

For each (k,τ) and $l < k$ there is a unique τ' such that

$$Q^k_\tau \subseteq Q^l_{\tau'},$$

$$\text{Diameter}(Q^k_\tau) \leq c_1\delta^k,$$

Each Q^k_τ contains some ball $B(z^k_\tau, a_0\delta^k)$.

For simplifying the discussion we may assume $\delta = \frac{1}{2}$ in Theorem 3.22. We say that a cube $Q \subset X$ is a dyadic cube if $Q = Q^k_\tau$ for some $k \in \mathbb{Z}$ and $\tau \in I_k$. The index τ indicates the position of these cubes. Theorem 3.22 says that the "length" of the "dyadic cube" Q^k_τ is roughly $\delta^k, 0 < \delta < 1$. For every integer j, we denote by $Q^{k,\nu}_\tau$, $\nu = 1, 2, ..., N(k,\tau)$ the dyadic cubes $Q^{k+j}_{\tau'} \subset Q^k_\tau$ and by $y^{k,\nu}_\tau$ any points in $Q^{k,\nu}_\tau$. In other words the cube $Q^{k,\nu}_{\tau'}$ are all the cubes contained in Q^k_τ with the side length δ^{k+j}. Here $Q^{k,\nu}_\tau$ should be denoted by $Q^{k,\nu}_{\tau'}$ because τ and τ' are different. The role of j is to refine the partition of X by the dyadic cubes Q^k_τ together with the accuracy of the approximation of integrals by a Riemann sum. Suppose that $D_k(x,y)$ are mother functions. By Coifman's decomposition of the identity operator, for each $f \in L^2(X)$,

$$f(x) = T_N(f)(x) + R_N(f)(x) = \sum_k D^N_k D_k(f)(x) + R_N(f)(x)$$

where $D^N_k(x,y)$, the kernel of D^N_k, are wavelets as defined in the Introduction.

We now replace

$$\sum_k D^N_k D_k(f)(x) = \sum_k \int_X D^N_k(x,y) D_k(f)(y) d\mu(y)$$

by a discrete version, obtained by paving X into the union of the dyadic cubes $Q^{k,\nu}_\tau$, where for each fixed $k \in \mathbb{Z}$ the union takes over $\tau \in I_k$ and $\nu, 1 \leq \nu \leq N(k,\tau)$. This leads to the following approximation

$$f(x) \approx \sum_{k\in\mathbb{Z}} \sum_{\tau\in I_k} \sum_{\nu=1}^{N(k,\tau)} \mu(Q^{k,\nu}_\tau) D^N_k(x, y^{k,\nu}_\tau) D_k(f)(y^{k,\nu}_\tau).$$

We define the operator $S = S_{N,j}$ by

$$S(f)(x) = \sum_{k\in\mathbb{Z}} \sum_{\tau\in I_k} \sum_{\nu=1}^{N(k,\tau)} \mu(Q^{k,\nu}_\tau) D^N_k(x, y^{k,\nu}_\tau) D_k(f)(y^{k,\nu}_\tau),$$

and the operator $\widetilde{R}_N = \widetilde{R}_{N,j}$ by

$$\widetilde{R}_N(f)(x) = \sum_{k\in\mathbb{Z}} \sum_{\tau\in I_k} \sum_{\nu=1}^{N(k,\tau)} \int_{Q_\tau^{k,\nu}} [D_k^N(x,y)D_k(f)(y) - D_k^N(x,y_\tau^{k,\nu})D_k(f)(y_\tau^{k,\nu})]d\mu(y).$$

Then we have a discrete version of decomposition on the identity operator:

$$I = S + \widetilde{R}_N + R_N.$$

The L^2-boundedness of S follows easily from the theory of frames described in Section 3.2. Indeed, for any fixed N, we write $\psi_\lambda(x) = \sqrt{\mu(Q_\tau^{k,\nu})}D_k(y_\tau^{k,\nu},x)$, where $\lambda \in \Lambda$, the set $\Lambda = \{(k,\tau,\nu) : k \in \mathbb{Z}, \tau \in I_k, 1 \le \nu \le N(k,\tau)\}$ and $\psi_\lambda^N(x) = \sqrt{\mu(Q_\tau^{k,\nu})}D_k^N(x,y_\tau^{k,\nu})$. Then the L^2-boundedness of S follows from the fact that both ψ_λ and ψ_λ^N are almost orthogonal in $L^2(X)$ as in the Definition 3.2. We prove this fact only for ψ_λ because the same proof applies to ψ_λ^N. We write

$$\begin{aligned}
&\left\| \sum_{\lambda\in\Lambda} \alpha_\lambda \psi_\lambda(x) \right\|_2^2 \\
&= \int \sum_{k\in\mathbb{Z}} \sum_{\tau\in I_k} \sum_{\nu=1}^{N(k,\tau)} (\mu(Q_\tau^{k,\nu}))^{\frac{1}{2}} \alpha_\tau^{k,\nu} D_k(y_\tau^{k,\nu},x) \\
&\quad \times \sum_{k'\in\mathbb{Z}} \sum_{\tau'\in I_{k'}} \sum_{\nu'=1}^{N(k',\tau')} (\mu(Q_{\tau'}^{k',\nu'}))^{\frac{1}{2}} \alpha_{\tau'}^{k',\nu'} D_{k'}(y_{\tau'}^{k',\nu'},x) d\mu(x) \\
&= \sum_{k\in\mathbb{Z}} \sum_{\tau\in I_k} \sum_{\nu=1}^{N(k,\tau)} \sum_{k'\in\mathbb{Z}} \sum_{\tau'\in I_{k'}} \sum_{\nu'=1}^{N(k',\tau')} \alpha_\tau^{k,\nu} \alpha_{\tau'}^{k',\nu'} I(k,\tau,\nu,k',\tau',\nu')
\end{aligned}$$

where

$$I(k,\tau,\nu,k',\tau',\nu') = \int (\mu(Q_\tau^{k,\nu}))^{\frac{1}{2}} (\mu(Q_{\tau'}^{k',\nu'}))^{\frac{1}{2}} D_k(y_\tau^{k,\nu},x) D_{k'}(y_{\tau'}^{k',\nu'},x) d\mu(x).$$

By the almost orthogonal argument for D_k, we have

$$\begin{aligned}
&\left| \int D_k(y_\tau^{k,\nu},x) D_{k'}(y_{\tau'}^{k',\nu'},x) d\mu(x) \right| \\
&\quad \le C 2^{-|k-k'|\epsilon'} \frac{2^{-(k\wedge k')\epsilon}}{(2^{-(k\wedge k')} + \rho(y_\tau^{k,\nu}, y_{\tau'}^{k',\nu'}))^{1+\epsilon}} \\
&\quad \le C 2^{(k\wedge k')} 2^{-|k-k'|\epsilon'} \left[\frac{2^{-(k\wedge k')}}{(2^{-(k\wedge k')} + \rho(y_\tau^{k,\nu}, y_{\tau'}^{k',\nu'}))} \right]^{(1+\epsilon)}
\end{aligned}$$

where $0 < \epsilon' < \epsilon$. Thus,

$$|I(k, \tau, \nu, k', \tau', \nu')| \leq C2^{-|k-k'|(\epsilon'+\frac{1}{2})} \left[\frac{2^{-(k\wedge k')}}{(2^{-(k\wedge k')} + \rho(y_\tau^{k,\nu}, y_{\tau'}^{k',\nu'}))} \right]^{(1+\epsilon)}.$$

The above estimate together with Schur's lemma gives the desired conclusion. Schur's lemma says the following. If one is given a matrix $M = m(x, y)$, $x \in E, y \in E$, (E is a set) and if there exists a positive and finite weight function $\omega(x)$ such that $\sum_{x\in E} |m(x, y)|\omega(x) \leq \omega(y)$ for every y and $\sum_{y\in E} |m(x, y)|\omega(y) \leq \omega(x)$ for every x, then the operator norm of M acting on $l^2(E)$ does not exceed 1.

We now prove that the operator S is invertible on $L^2(X)$. Indeed, we will further prove that S^{-1}, the inverse of S on $L^2(X)$, also maps $\mathcal{M}_0(x_1, r, \beta, \gamma)$ into itself for $x_1 \in X, r > 0$ and $0 < \beta, \gamma < \epsilon$. As a consequence of theses results, we provide wavelet expansions on spaces of homogeneous type. To this end, note that $I = S + \widetilde{R}_N + R_N$. We have proved in Section 3.4 that the operator norm of R_N on $L^p, 1 < p < \infty$, and $\mathcal{M}_0(x_1, r, \beta, \gamma)$ is at most $C2^{-N\delta}$ for a constant $C > 0$ and some $\delta > 0$. The following result gives the estimate on the operator $\widetilde{R}_N$. In this lemma the integer j is any fixed integer.

Lemma 3.23. *The operator $\widetilde{R}_N$ is defined as above and depends on the integer $j \geq 1$ which is used to improve the accuracy in the Riemann sum defining $\widetilde{R}_N$. Then the kernel $\widetilde{R}_N(x, y)$, of $\widetilde{R}_N$, satisfies the following estimates: for $0 < \epsilon' < \epsilon$ there exist a constants $0 < C < \infty$ and $\delta > 0$ such that for integer $N > 0$,*

$$|\widetilde{R}_N(x, y)| \leq C2^{-j\delta}\rho(x, y)^{-1}; \tag{3.70}$$

$$|\widetilde{R}_N(x, y) - \widetilde{R}_N(x, y')| \leq C2^{-j\delta}\rho(y, y')^{\epsilon'}\rho(x, y)^{-(1+\epsilon')} \tag{3.71}$$

for $\rho(y, y') \leq \frac{1}{2A}\rho(x, y)$;

$$|\widetilde{R}_N(x, y) - \widetilde{R}_N(x', y)| \leq C2^{-j\delta}\rho(x, x')^{\epsilon'}\rho(x, y)^{-(1+\epsilon')} \tag{3.72}$$

for $\rho(x, x') \leq \frac{1}{2A}\rho(x, y)$;

$$\begin{aligned} &|\widetilde{R}_N(x, y) - \widetilde{R}_N(x', y) - \widetilde{R}_N(x, y') + \widetilde{R}_N(x', y')| \\ &\quad \leq C2^{-j\delta}\rho(x, x')^{\epsilon'}\rho(y, y')^{\epsilon'}\rho(x, y)^{-(1+2\epsilon')} \end{aligned} \tag{3.73}$$

for $\rho(x, x') \leq \frac{1}{2A}\rho(x, y)$ and $\rho(y, y') \leq \frac{1}{2A}\rho(x, y)$;

$$|\langle \widetilde{R}_N, f\rangle| \leq C2^{-j\delta}r \tag{3.74}$$

for all $f \in C_0^\eta(X \times X)$ with $\mathrm{supp}(f) \subseteq B(x_1, r) \times B(y_1, r), x_1$ and $y_1 \in X$, $\|f\|_\infty \leq 1$, $\|f(\cdot, y)\|_\eta \leq r^{-\eta}$, and $\|f(x, \cdot)\|_\eta \leq r^{-\eta}$ for all x and $y \in X$.

We point out that the almost orthogonal estimate was used in the proof of Lemma 3.17. But, however, this almost orthogonal argument is not needed for the proof of Lemma 3.23. The key facts used in the proof of Lemma 3.23 are the smoothness condition of $D_k(x,y)$ on the variable y and the fact that the side length of $Q_\tau^{k,\nu}$ is equivalent to 2^{-k-j}.

We now prove Lemma 3.23. First of all, we rewrite $\widetilde{R}_N$ as follows.

$$\begin{aligned}\widetilde{R}_N(f)(x) =& \sum_{k\in\mathbb{Z}}\sum_{\tau\in I_k}\sum_{\nu=1}^{N(k,\tau)}\int_{Q_\tau^{k,\nu}}[D_k^N(x,y)-D_k^N(x,y_\tau^{k,\nu})]D_k(f)(y)d\mu(y) \quad (3.75)\\ &+\sum_{k\in\mathbb{Z}}\sum_{\tau\in I_k}\sum_{\nu=1}^{N(k,\tau)}\int_{Q_\tau^{k,\nu}}D_k^N(x,y_\tau^{k,\nu})[D_k(f)(y)-D_k(f)(y_\tau^{k,\nu})]d\mu(y)\\ =& R_1(f)(x)+R_2(f)(x).\end{aligned}$$

We prove that the kernels of R_1 and R_2 satisfy the estimates of (3.70) – (3.74). It is easy to see that if the kernel of R_1 is denoted by $R_1(x,y)$, then

$$R_1(x,y)=\sum_{k\in\mathbb{Z}}\sum_{\tau\in I_k}\sum_{\nu=1}^{N(k,\tau)}\int_{Q_\tau^{k,\nu}}[D_k^N(x,z)-D_k^N(x,y_\tau^{k,\nu})]D_k(z,y)d\mu(z).$$

Note that $D_k^N(x,y)$, the kernel of D_k^N, satisfies the same estimates (3.14) – (3.19) with the constant C depending on N, and observe that the side length of $Q_\tau^{k,\nu}$ is equivalent to 2^{-k-j}. We obtain

$$\begin{aligned}&\int_{Q_\tau^{k,\nu}}|D_k^N(x,z)-D_k^N(x,y_\tau^{k,\nu})||D_k(z,y)|d\mu(z)\\ &\quad\le C\int_{Q_\tau^{k,\nu}}\left(\frac{\rho(z,y_\tau^{k,\nu})}{2^{-k}+\rho(x,z)}\right)^\epsilon\frac{2^{-k\epsilon}}{(2^{-k}+\rho(x,z))^{1+\epsilon}}|D_k(z,y)|d\mu(z)\\ &\quad\le C2^{-j\epsilon}\int_{Q_\tau^{k,\nu}}\frac{2^{-k\epsilon}}{(2^{-k}+\rho(x,z))^{1+\epsilon}}|D_k(z,y)|d\mu(z).\end{aligned}$$

Thus,

$$\begin{aligned}|R_1(x,y)|&\le C2^{-j\epsilon}\sum_{k\in\mathbb{Z}}\sum_{\tau\in I_k}\sum_{\nu=1}^{N(k,\tau)}\int_{Q_\tau^{k,\nu}}\frac{2^{-k\epsilon}}{(2^{-k}+\rho(x,z))^{1+\epsilon}}|D_k(z,y)|d\mu(z)\\ &\le C2^{-j\epsilon}\sum_{k\in\mathbb{Z}}\int_X\frac{2^{-k\epsilon}}{(2^{-k}+\rho(x,z))^{1+\epsilon}}|D_k(z,y)|d\mu(z)\\ &\le C2^{-j\epsilon}\sum_{k\in\mathbb{Z}}\frac{2^{-k\epsilon}}{(2^{-k}+\rho(x,y))^{1+\epsilon}}\le C2^{-j\epsilon}\rho(x,y)^{-1} \quad (3.76)\end{aligned}$$

which implies that the kernel of R_1 satisfies the estimate (3.70) with the constant $C2^{-j\epsilon}$.

To show (3.71) for R_1, we write

$$R_1(x,y) - R_1(x,y') = \sum_{k\in\mathbb{Z}} \sum_{\tau\in I_k} \sum_{\nu=1}^{N(k,\tau)} \int_{Q_\tau^{k,\nu}} [D_k^N(x,z) - D_k^N(x,y_\tau^{k,\nu})][D_k(z,y) - D_k(z,y')]d\mu(z).$$

For $z \in X$, consider two cases: (i) $W_1 = \{z \in X : \rho(y,y') \le \frac{1}{2A}(2^{-k} + \rho(z,y))\}$ and (ii) $W_2 = \{z \in X : \frac{1}{4A^2}\rho(x,y) \ge \rho(y,y') \ge \frac{1}{2A}(2^{-k} + \rho(z,y))\}$. For the first case,

$$|D_k^N(x,z) - D_k^N(x,y_\tau^{k,\nu})||D_k(z,y) - D_k(z,y')| \le C\frac{2^{-k\epsilon}}{(2^{-k}+\rho(x,z))^{1+\epsilon}}\left(\frac{\rho(y,y')}{2^{-k}+\rho(z,y)}\right)^\epsilon \frac{2^{-k\epsilon}}{(2^{-k}+\rho(z,y))^{1+\epsilon}}$$

where $y_\tau^{k,\nu} \in Q_\tau^{k,\nu}$.

For the second case, note that $\frac{1}{4A^2}\rho(x,y) \ge \rho(y,y') \ge \frac{1}{2A}(2^{-k} + \rho(z,y))$ implies $\rho(y,y') \ge \frac{1}{A}2^{-1-k}$ and $\frac{1}{2A}\rho(x,y) \le \rho(x,z)$. This yields

$$|D_k^N(x,z) - D_k^N(x,y_\tau^{k,\nu})||D_k(z,y) - D_k(z,y')| \le C\frac{2^{-k\epsilon}}{(2^{-k}+\rho(x,z))^{1+\epsilon}}\left[\frac{2^{-k\epsilon}}{(2^{-k}+\rho(z,y))^{1+\epsilon}} + \frac{2^{-k\epsilon}}{(2^{-k}+\rho(z,y'))^{1+\epsilon}}\right].$$

Thus, if $\frac{1}{4A^2}\rho(x,y) \ge \rho(y,y')$,

$$\begin{aligned} &|R_1(x,y) - R_1(x,y')| \\ &\le C\sum_{k\in\mathbb{Z}} \int_{W_1} \frac{2^{-k\epsilon}}{(2^{-k}+\rho(x,z))^{1+\epsilon}}\left(\frac{\rho(y,y')}{2^{-k}+\rho(z,y)}\right)^\epsilon \frac{2^{-k\epsilon}}{(2^{-k}+\rho(z,y))^{1+\epsilon}}d\mu(z) \\ &+C\sum_{k\in\mathbb{Z}} \int_{W_2} \frac{2^{-k\epsilon}}{(2^{-k}+\rho(x,z))^{1+\epsilon}}\left[\frac{2^{-k\epsilon}}{(2^{-k}+\rho(z,y))^{1+\epsilon}} + \frac{2^{-k\epsilon}}{(2^{-k}+\rho(z,y'))^{1+\epsilon}}\right]d\mu(z) \\ &\le C\sum_{k\in\mathbb{Z}}\left(\frac{\rho(y,y')}{2^{-k}+\rho(x,y)}\right)^{\epsilon'} \frac{2^{-k(\epsilon-\epsilon')}}{(2^{-k}+\rho(x,y))^{1+(\epsilon-\epsilon')}} \le C\rho(y,y')^{\epsilon'}\rho(x,y)^{-(1+\epsilon')} \end{aligned} \tag{3.77}$$

where $0 < \epsilon' < \epsilon$.

The estimate (3.76) yields that if $\frac{1}{4A^2}\rho(x,y) \ge \rho(y,y')$,

$$|R_1(x,y) - R_1(x,y')| \le C2^{-j\epsilon}\rho(x,y)^{-1}. \tag{3.78}$$

Taking the geometrical mean of (3.77) and (3.78) implies that $R_1(x,y)$, the kernel of R_1, satisfies (3.71) with the constant $C2^{-j\delta}$.

Similarly, considering two cases: (i) $W_1 = \{z \in X : \rho(x,x') \leq \frac{1}{2A}(2^{-k} + \rho(x,z))\}$ and (ii) $W_2 = \{z \in X : \frac{1}{4A^2}\rho(x,y) \geq \rho(x,x') \geq \frac{1}{2A}(2^{-k} + \rho(x,z))\}$, then we have

$$
\begin{aligned}
&|R_1(x,y) - R_1(x',y)| \qquad (3.79)\\
&\leq \sum_{k\in\mathbb{Z}} \sum_{\tau\in I_k} \sum_{\nu=1}^{N(k,\tau)} \int_{Q_\tau^{k,\nu}} |[D_k^N(x,z) - D_k^N(x,y_\tau^{k,\nu})] \\
&\quad -[D_k^N(x',z) - D_k^N(x',y_\tau^{k,\nu})]||D_k(z,y)|d\mu(z)\\
&\leq C\sum_{k\in\mathbb{Z}} \int_{W_1} \frac{2^{-k\epsilon}}{(2^{-k}+\rho(x,z))^{1+\epsilon}} \left(\frac{\rho(x,x')}{2^{-k}+\rho(x,z)}\right)^{\epsilon} \frac{2^{-k\epsilon}}{(2^{-k}+\rho(z,y))^{1+\epsilon}} d\mu(z)\\
&\quad +C\sum_{k\in\mathbb{Z}} \int_{W_2} \left[\frac{2^{-k\epsilon}}{(2^{-k}+\rho(x,z))^{1+\epsilon}} + \frac{2^{-k\epsilon}}{(2^{-k}+\rho(x',z))^{1+\epsilon}}\right]\\
&\quad \times \frac{2^{-k\epsilon}}{(2^{-k}+\rho(z,y))^{1+\epsilon}} d\mu(z)\\
&\leq C\sum_{k\in\mathbb{Z}} \left(\frac{\rho(x,x')}{2^{-k}+\rho(x,y)}\right)^{\epsilon'} \frac{2^{-k(\epsilon-\epsilon')}}{(2^{-k}+\rho(x,y))^{1+(\epsilon-\epsilon')}}\\
&\leq C\rho(x,x')^{\epsilon'}\rho(x,y)^{-(1+\epsilon')}
\end{aligned}
$$

where $0 < \epsilon' < \epsilon$.

As above, the estimate (3.79) together with the estimate (3.76) implies that $R_1(x,y)$ satisfies the estimate (3.72) with the constant $C2^{-j\delta}$.

We now prove the estimate (3.73) for $R_1(x,y)$. We first write

$$
\begin{aligned}
&|R_1(x,y) - R_1(x',y) - R_1(x,y') + R_1(x',y')|\\
&\leq \sum_{k\in\mathbb{Z}} \sum_{\tau\in I_k} \sum_{\nu=1}^{N(k,\tau)} \int_{Q_\tau^{k,\nu}} \Big|[D_k^N(x,z) - D_k^N(x,y_\tau^{k,\nu})] \\
&\quad -[D_k^N(x',z) - D_k^N(x',y_\tau^{k,\nu})]\Big| \times |D_k(z,y) - D_k(z,y')|d\mu(z).
\end{aligned}
$$

If $\frac{1}{4A^2}\rho(x,y) \geq \rho(x,x')$ and $\frac{1}{4A^2}\rho(x,y) \geq \rho(y,y')$, then for any $z \in X$, we have the following three cases: (i) $W_1 = \{z \in X : \rho(x,x') \leq \frac{1}{2A}(2^{-k} + \rho(x,z))$ and $\rho(y,y') \leq \frac{1}{2A}(2^{-k} + \rho(z,y))\}$; (ii) $W_2 = \{z \in X : \rho(x,x') \leq \frac{1}{2A}(2^{-k} + \rho(x,z))$ and $\rho(y,y') \geq \frac{1}{2A}(2^{-k} + \rho(z,y))\}$; and (iii) $W_3 = \{z \in X : \rho(x,x') \geq \frac{1}{2A}(2^{-k} + \rho(x,z))$ and $\rho(y,y') \leq \frac{1}{2A}(2^{-k} + \rho(z,y))\}$.

If $z \in W_1 \cap Q_\tau^{k,\nu}$, then

$$
\begin{aligned}
&|[D_k^N(x,z) - D_k^N(x,y_\tau^{k,\nu})] - [D_k^N(x',z) - D_k^N(x,y_\tau^{k,\nu})]||D_k(z,y) - D_k(z,y')| \\
&\quad \le C\bigg(\frac{\rho(x,x')}{2^{-k}+\rho(x,z)}\bigg)^\epsilon \frac{2^{-k\epsilon}}{(2^{-k}+\rho(x,z))^{1+\epsilon}} \bigg(\frac{\rho(y,y')}{2^{-k}+\rho(z,y)}\bigg)^\epsilon \\
&\qquad \times \frac{2^{-k\epsilon}}{(2^{-k}+\rho(z,y))^{1+\epsilon}}.
\end{aligned}
$$

If $z \in W_2 \cap Q_\tau^{k,\nu}$, then $\rho(y,y') \ge \frac{1}{A}2^{-1-k}$ and $\frac{1}{2A}\rho(x,y) \le \rho(x,z)$. Thus,

$$
\begin{aligned}
&|[D_k^N(x,z) - D_k^N(x,y_\tau^{k,\nu})] - [D_k^N(x',z) - D_k^N(x',y_\tau^{k,\nu})]||D_k(z,y) - D_k(z,y')| \\
&\le C\bigg(\frac{\rho(x,x')}{2^{-k}+\rho(x,z)}\bigg)^\epsilon \frac{2^{-k\epsilon}}{(2^{-k}+\rho(x,z))^{1+\epsilon}} \\
&\qquad \times \bigg\{\frac{2^{-k\epsilon}}{(2^{-k}+\rho(z,y))^{1+\epsilon}} + \frac{2^{-k\epsilon}}{(2^{-k}+\rho(z,y'))^{1+\epsilon}}\bigg\} \\
&\le C\bigg(\frac{\rho(x,x')}{2^{-k}+\rho(x,y)}\bigg)^\epsilon \frac{2^{-k\epsilon}}{(2^{-k}+\rho(x,y))^{1+\epsilon}} \\
&\qquad \times \bigg\{\frac{2^{-k\epsilon}}{(2^{-k}+\rho(z,y))^{1+\epsilon}} + \frac{2^{-k\epsilon}}{(2^{-k}+\rho(z,y'))^{1+\epsilon}}\bigg\}.
\end{aligned}
$$

If $z \in W_3 \cap Q_\tau^{k,\nu}$, then $\rho(x,x') \ge \frac{1}{A}2^{-1-k}$ and $\frac{1}{2A}\rho(x,y) \le \rho(z,y)$ yield

$$
\begin{aligned}
&|[D_k^N(x,z) - D_k^N(x,y_\tau^{k,\nu})] - [D_k^N(x',z) - D_k^N(x',y_\tau^{k,\nu})]||D_k(z,y) - D_k(z,y')| \\
&\le C\bigg[\frac{2^{-k\epsilon}}{(2^{-k}+\rho(x,z))^{1+\epsilon}} + \frac{2^{-k\epsilon}}{(2^{-k}+\rho(x',z))^{1+\epsilon}}\bigg] \bigg(\frac{\rho(y,y')}{2^{-k}+\rho(z,y)}\bigg)^\epsilon \\
&\qquad \times \frac{2^{-k\epsilon}}{(2^{-k}+\rho(z,y))^{1+\epsilon}} \\
&\le C\bigg[\frac{2^{-k\epsilon}}{(2^{-k}+\rho(x,z))^{1+\epsilon}} + \frac{2^{-k\epsilon}}{(2^{-k}+\rho(x',z))^{1+\epsilon}}\bigg] \bigg(\frac{\rho(y,y')}{2^{-k}+\rho(x,y)}\bigg)^\epsilon \\
&\qquad \times \frac{2^{-k\epsilon}}{(2^{-k}+\rho(x,y))^{1+\epsilon}}.
\end{aligned}
$$

We now get

$$
\begin{aligned}
&|R_1(x,y) - R_1(x',y) - R_1(x,y') + R_1(x',y')| \\
&\quad \le C\sum_{k\in\mathbb{Z}} \int_{W_1} \bigg(\frac{\rho(x,x')}{2^{-k}+\rho(x,z)}\bigg)^\epsilon \frac{2^{-k\epsilon}}{(2^{-k}+\rho(x,z))^{1+\epsilon}} \\
&\qquad\qquad \times \bigg(\frac{\rho(y,y')}{2^{-k}+\rho(z,y)}\bigg)^\epsilon \frac{2^{-k\epsilon}}{(2^{-k}+\rho(z,y))^{1+\epsilon}} d\mu(z)
\end{aligned}
\tag{3.80}
$$

$$+C\sum_{k\in\mathbb{Z}}\int_{W_2}\left(\frac{\rho(x,x')}{2^{-k}+\rho(x,y)}\right)^{\epsilon}\frac{2^{-k\epsilon}}{(2^{-k}+\rho(x,y))^{1+\epsilon}}$$
$$\times\left\{\frac{2^{-k\epsilon}}{(2^{-k}+\rho(z,y))^{1+\epsilon}}+\frac{2^{-k\epsilon}}{(2^{-k}+\rho(z,y'))^{1+\epsilon}}\right\}d\mu(z)$$
$$+C\sum_{k\in\mathbb{Z}}\int_{W_3}\left[\frac{2^{-k\epsilon}}{(2^{-k}+\rho(x,z))^{1+\epsilon}}+\frac{2^{-k\epsilon}}{(2^{-k}+\rho(x',z))^{1+\epsilon}}\right]$$
$$\times\left(\frac{\rho(y,y')}{2^{-k}+\rho(x,y)}\right)^{\epsilon}\frac{2^{-k\epsilon}}{(2^{-k}+\rho(x,y))^{1+\epsilon}}d\mu(z)$$
$$\leq C\rho(x,x')^{\epsilon'}\rho(y,y')^{\epsilon'}\rho(x,y)^{-(1+2\epsilon')}$$

where $0<\epsilon'<\epsilon$.

On the other hand, if $\frac{1}{4A^2}\rho(x,y)\geq\rho(x,x')$ and $\frac{1}{4A^2}\rho(x,y)\geq\rho(y,y')$, then, by the estimate (3.76), we have

$$\begin{aligned}|R_1(x,y)-R_1(x',y)-R_1(x,y')+R_1(x',y')|\\ \leq|R_1(x,y)|+|R_1(x',y)|+|R_1(x,y')|+|R_1(x',y')|\\ \leq C2^{-j\epsilon}\rho(x,y)^{-1}.\end{aligned} \tag{3.81}$$

Again, as above, the geometrical mean of (3.80) and (3.81) implies that $R_1(x,y)$ satisfies the estimate (3.73).

The proof of (3.74) for R_1 is same as in the proof of Lemma 3.17. The proofs for R_2 are similar to the proofs of R_1. We leave these details to the reader. By Lemma 3.17, as mentioned above, R_N satisfies all estimates of (3.70) - (3.74).

Note that $\widetilde{R}_N(1)=\widetilde{R}_N^*(1)=0$ and the fact that $S^{-1}=\sum_{m=0}^{\infty}(\widetilde{R}_N+R_N)^m$. As an immediate consequence of Theorem 1.17 and Theorem 2.4, we obtain the following result.

Proposition 3.24 *If j and N are large integers, then S^{-1}, the inverse of the operator S, maps test function to test function, and it is bounded on L^p for $1<p<\infty$. Moreover, there exist constant C which is independent of $f\in\mathcal{M}_0(x_1,r,\beta,\gamma)$ with $x_1\in X$, $r>0$ and $0<\beta,\ \gamma<\epsilon$ and C_p which depends only on p for $1<p<\infty$, such that*

$$\|S^{-1}(f)\|_{\mathcal{M}(x_1,r,\beta,\gamma)}\leq C\|f\|_{\mathcal{M}(x_1,r,\beta,\gamma)}, \tag{3.82}$$

and

$$\|S^{-1}(f)\|_p\leq C_p\|f\|_p. \tag{3.83}$$

We now prove the main result, namely wavelet expansions on spaces of homogeneous type ([H3]). Wavelets are defined in Definition 2.2.

Theorem 3.25. *There exist three families* $\psi_\lambda(x), \widetilde{\psi}_\lambda(x), \widetilde{\widetilde{\psi}}_\lambda(x)$ *of wavelets such that, for every square integrable function* f *one has*

$$f(x) = \sum_{\lambda \in \Lambda} \widetilde{\psi}_\lambda(x)\langle f, \psi_\lambda\rangle \tag{3.84}$$
$$= \sum_{\lambda \in \Lambda} \psi_\lambda(x)\langle f, \widetilde{\widetilde{\psi}}_\lambda\rangle.$$

The first family is given by

$$\psi_\lambda(x) = \sqrt{\mu(Q_\tau^{k,\nu})} D_k(y_\tau^{k,\nu}, x), \ \ \lambda \in \Lambda$$
$$= \{(k,\tau,\nu) : k \in \mathbb{Z}, \tau \in I_k, 1 \leq \nu \leq N(k,\tau)\}$$

where $D_k(x,y)$ *are mother functions and the sampling grid* $y_\tau^{k,\nu}$ *is given by any point* $y_\tau^{k,\nu} \in Q_\tau^{k,\nu}$.

The second and third family have a similar definition.

Let $\widetilde{\psi}_\lambda(x) = S^{-1}(D_k^N(\cdot, y_\tau^{k,\nu})(x)$ where $D_k^N(x,y)$ are kernels of D_k^N and N is a fixed large integer. The Proposition 3.24 implies that $\widetilde{D}_k(x,y)$ satisfy the estimates of (3.14) and (3.15) with ϵ replaced by ϵ', for $0 < \epsilon' < \epsilon$, and $\int \widetilde{D}_k(x,y)d\mu(y) = \int \widetilde{D}_k(x,y)d\mu(x) = 0$ for all $k \in \mathbb{Z}$, follow from the facts that $D_k^N(1) = (S^{-1})^*(1) = 0$. Similarly, let $\widetilde{\widetilde{\psi}}_\lambda(x) = D_k^N S^{-1}(\cdot, y_\tau^{k,\nu})(x)$. Then Proposition 3.24 implies that $\widetilde{\widetilde{D}}_k(x,y)$ satisfy the estimates of (3.14) and (3.16) with ϵ replaced by ϵ', for $0 < \epsilon' < \epsilon$. $\int \widetilde{\widetilde{D}}_k(x,y)d\mu(y) = \int \widetilde{\widetilde{D}}_k(x,y)d\mu(x) = 0$ for all $k \in Z$, follow from the facts that $(D_k^N)^*(1) = (S^{-1})(1) = 0$. These facts yield that $\widetilde{\psi}_\lambda$ and $\widetilde{\widetilde{\psi}}_\lambda$ are wavelets. It remains to prove that the series in (3.84) converge in $L^2(X)$. Since

$$\sum_{|k|\leq M} \sum_{\tau \in I_k} \sum_{\nu=1}^{N(k,\tau)} \mu(Q_\tau^{k,\nu})\widetilde{D}_k(x, y_\tau^{k,\nu})D_k(f)(y_\tau^{k,\nu})$$
$$= S^{-1}\Big\{ \sum_{|k|\leq M} \sum_{\tau \in I_k} \sum_{\nu=1}^{N(k,\tau)} \mu(Q_\tau^{k,\nu})D_k^N(\cdot, y_\tau^{k,\nu})D_k(f)(y_\tau^{k,\nu})\Big\}(x)$$
$$= S^{-1}\Big\{ S(f)(\cdot) - \sum_{|k|\geq M} \sum_{\tau \in I_k} \sum_{\nu=1}^{N(k,\tau)} \mu(Q_\tau^{k,\nu})D_k^N(\cdot, y_\tau^{k,\nu})D_k(f)(y_\tau^{k,\nu})\Big\}(x)$$
$$= f(x) - \lim_{m\to\infty} (R)^m(f)(x) - S^{-1}$$
$$\times \Big\{ \sum_{|k|\geq M} \sum_{\tau \in I_k} \sum_{\nu=1}^{N(k,\tau)} \mu(Q_\tau^{k,\nu})D_k^N(\cdot, y_\tau^{k,\nu})D_k(f)(y_\tau^{k,\nu})\Big\}(x)$$

where $R = \widetilde{R}_N + R_N$.

So all we need to do is to show that $(R)^m(f)(x)$ and

$$S^{-1}\Big\{\sum_{|k|\geq M}\sum_{\tau\in I_k}\sum_{\nu=1}^{N(k,\tau)}\mu(Q_\tau^{k,\nu})D_k^N(\cdot,y_\tau^{k,\nu})D_k(f)(y_\tau^{k,\nu})\Big\}(x)$$

converge to zero in $L^2(X)$ as m, M tend to infinity.

By the fact $\|(R)^m(f)\|_2 \leq C^m(2^{-j\delta}+2^{-N\delta})^m\|f\|_2$, it follows that $\lim_{m\to\infty}(R)^m(f)(x) = 0$ in the norm of L^2, for fixed large integers j and N. To show

$$S^{-1}\Big\{\sum_{|k|\geq M}\sum_{\tau\in I_k}\sum_{\nu=1}^{N(k,\tau)}\mu(Q_\tau^{k,\nu})D_k^N(\cdot,y_\tau^{k,\nu})D_k(f)(y_\tau^{k,\nu})\Big\}(x)$$

converges to zero in the norm of $L^2(X)$ as M tends to infinity, it suffices to prove that

$$\lim_{M\to\infty}\Big\|\sum_{|k|\geq M}\sum_{\tau\in I_k}\sum_{\nu=1}^{N(k,\tau)}\mu(Q_\tau^{k,\nu})D_k^N(x,y_\tau^{k,\nu})D_k(f)(y_\tau^{k,\nu})\Big\|_2 = 0. \quad (3.85)$$

This estimate will follow from the general situation which we describe now. Indeed the wavelet expansions in Theorem 3.25 still hold on $\mathcal{M}(\beta',\gamma')$ and $L^p, 1 < p < \infty$. To see these results, all we need to do is to show that if $f \in \mathcal{M}_0(\beta,\gamma), (R)^m(f)(x)$ and

$$S^{-1}\Big\{\sum_{|k|\geq M}\sum_{\tau\in I_k}\sum_{\nu=1}^{N(k,\tau)}\mu(Q_\tau^{k,\nu})D_k^N(\cdot,y_\tau^{k,\nu})D_k(f)(y_\tau^{k,\nu})\Big\}(x)$$

converge to zero in the norm of $\mathcal{M}(\beta',\gamma')$ for $0 < \beta' < \beta$ and $0 < \gamma' < \gamma$ and $L^p(X)$, for $1 < p < \infty$, as m, M tend to infinity.

By Theorem 1.10, for $1 < p < \infty$,

$$\|(R)^m(f)\|_p \leq C^m(2^{-j\delta}+2^{-N\delta})^m\|f\|_p$$

which implies $\lim_{m\to\infty}(R)^m(f)(x) = 0$ in the norm of $L^p, 1 < p < \infty$, as m tends to infinity uniformly for large integers j and N. To show

$$S^{-1}\Big\{\sum_{|k|\geq M}\sum_{\tau\in I_k}\sum_{\nu=1}^{N(k,\tau)}\mu(Q_\tau^{k,\nu})D_k^N(\cdot,y_\tau^{k,\nu})D_k(f)(y_\tau^{k,\nu})\Big\}(x)$$

converges to zero in the norm of $L^p(X)$, for $1 < p < \infty$, as M tends to infinity, it suffices to show that for $1 < p < \infty$,

$$\lim_{M\to\infty}\left\|\sum_{|k|\geq M}\sum_{\tau\in I_k}\sum_{\nu=1}^{N(k,\tau)}\mu(Q_\tau^{k,\nu})D_k^N(x,y_\tau^{k,\nu})D_k(f)(y_\tau^{k,\nu})\right\|_p=0. \quad (3.86)$$

Let $\frac{1}{p}+\frac{1}{q}=1$. By the duality argument,

$$\begin{aligned}
&\left\|\sum_{|k|\geq M}\sum_{\tau\in I_k}\sum_{\nu=1}^{N(k,\tau)}\mu(Q_\tau^{k,\nu})D_k^N(x,y_\tau^{k,\nu})D_k(f)(y_\tau^{k,\nu})\right\|_p\\
&\quad=\sup_{\|g\|_q\leq 1}\left|\left\langle\sum_{|k|\geq M}\sum_{\tau\in I_k}\sum_{\nu=1}^{N(k,\tau)}\mu(Q_\tau^{k,\nu})D_k^N(x,y_\tau^{k,\nu})D_k(f)(y_\tau^{k,\nu}),g\right\rangle\right|\\
&\quad\leq\sup_{\|g\|_q\leq 1}\sum_{|k|\geq M}\sum_{\tau\in I_k}\sum_{\nu=1}^{N(k,\tau)}\mu(Q_\tau^{k,\nu})|(D_k^N)^*(g)(y_\tau^{k,\nu})||D_k(f)(y_\tau^{k,\nu})|\\
&\quad\leq\sup_{\|g\|_q\leq 1}\sum_{|k|\geq M}\sum_{\tau\in I_k}\sum_{\nu=1}^{N(k,\tau)}\int_{Q_\tau^{k,\nu}}|(D_k^N)^*(g)(y_\tau^{k,\nu})||D_k(f)(y_\tau^{k,\nu})|d\mu(y).
\end{aligned}$$

By a similar proof to Theorem 4.1, we obtain

$$\begin{aligned}
&\sum_{|k|\geq M}\sum_{\tau\in I_k}\sum_{\nu=1}^{N(k,\tau)}\int_{Q_\tau^{k,\nu}}|(D_k^N)^*(g)(y_\tau^{k,\nu})||D_k(f)(y_\tau^{k,\nu})|d\mu(y)\\
&\qquad\leq C\|g\|_q\left\|\left\{\sum_{|k|>M}|D_k(f)|^2\right\}^{\frac{1}{2}}\right\|_p.
\end{aligned}$$

Thus,

$$\begin{aligned}
&\left\|\sum_{|k|\geq M}\sum_{\tau\in I_k}\sum_{\nu=1}^{N(k,\tau)}\mu(Q_\tau^{k,\nu})D_k^N(x,y_\tau^{k,\nu})D_k(f)(y_\tau^{k,\nu})\right\|_p\\
&\qquad\leq C\left\|\left\{\sum_{|k|>M}|D_k(f)|^2\right\}^{\frac{1}{2}}\right\|_p
\end{aligned}$$

where the last term above, by the Littlewood-Paley estimates on $L^p, 1<p<\infty$, tends to zero as M tends to infinity.

We now prove that the series in (3.84) converge in the norm of $\mathcal{M}(\beta',\gamma')$ for each $f\in\mathcal{M}_0(\beta,\gamma)$ with $0<\beta'<\beta$, $0<\gamma'<\gamma$. This follows from the fact that for $0<\beta'<\beta$, $0<\gamma'<\gamma$ and $f\in\mathcal{M}_0(\beta,\gamma)$,

$$\lim_{M\to\infty}\left\|\sum_{|k|\geq M}\sum_{\tau\in I_k}\sum_{\nu=1}^{N(k,\tau)}\mu(Q_\tau^{k,\nu})D_k^N(x,y_\tau^{k,\nu})D_k(f)(y_\tau^{k,\nu})\right\|_{\mathcal{M}(\beta',\gamma')}=0. \quad (3.87)$$

To show (3.87), we denote

$$
\begin{aligned}
E_k(f)(x) &:= \sum_{\tau\in I_k}\sum_{\nu=1}^{N(k,\tau)} \mu(Q_\tau^{k,\nu}) D_k^N(x, y_\tau^{k,\nu}) D_k(f)(y_\tau^{k,\nu}) \\
&= \int \sum_{\tau\in I_k}\sum_{\nu=1}^{N(k,\tau)} \mu(Q_\tau^{k,\nu}) D_k^N(x, y_\tau^{k,\nu}) D_k(y_\tau^{k,\nu}, y) f(y) d\mu(y)
\end{aligned}
$$

where

$$
E_k(x,y) = \sum_{\tau\in I_k}\sum_{\nu=1}^{N(k,\tau)} \mu(Q_\tau^{k,\nu}) D_k^N(x, y_\tau^{k,\nu}) D_k(y_\tau^{k,\nu}, y).
$$

Then, $E_k(x,y)$, the kernel of E_k, satisfies the estimates of (3.14). To see this, we have

$$
\begin{aligned}
|E_k(x,y)| &= \left| \sum_{\tau\in I_k}\sum_{\nu=1}^{N(k,\tau)} \int_{Q_\tau^{k,\nu}} D_k^N(x, y_\tau^{k,\nu}) D_k(y_\tau^{k,\nu}, y) d\mu(z) \right| \\
&\le \sum_{\tau\in I_k}\sum_{\nu=1}^{N(k,\tau)} \int_{Q_\tau^{k,\nu}} |D_k^N(x, y_\tau^{k,\nu})||D_k(y_\tau^{k,\nu}, y)| d\mu(z) \\
&\le C\sum_{\tau\in I_k}\sum_{\nu=1}^{N(k,\tau)} \int_{Q_\tau^{k,\nu}} \frac{2^{-k\epsilon}}{(2^{-k}+\rho(x, y_\tau^{k,\nu}))^{1+\epsilon}} \frac{2^{-k\epsilon}}{(2^{-k}+\rho(y_\tau^{k,\nu}, y))^{1+\epsilon}} d\mu(z) \\
&\le C\sum_{\tau\in I_k}\sum_{\nu=1}^{N(k,\tau)} \int_{Q_\tau^{k,\nu}} \frac{2^{-k\epsilon}}{(2^{-k}+\rho(x, z))^{1+\epsilon}} \frac{2^{-k\epsilon}}{(2^{-k}+\rho(z, y))^{1+\epsilon}} d\mu(z) \\
&= C\int_X \frac{2^{-k\epsilon}}{(2^{-k}+\rho(x, z))^{1+\epsilon}} \frac{2^{-k\epsilon}}{(2^{-k}+\rho(z, y))^{1+\epsilon}} d\mu(z) \\
&\le C\frac{2^{-k\epsilon}}{(2^{-k}+\rho(x, y))^{1+\epsilon}}.
\end{aligned}
$$

Note that

$$
\begin{aligned}
&|D_k^N(x, y_\tau^{k,\nu}) - D_k^N(x', y_\tau^{k,\nu})| \\
&\le C\Big(\frac{\rho(x,x')}{2^{-k}+\rho(x, y_\tau^{k,\nu})}\Big)^{\epsilon} \Big[\frac{2^{-k\epsilon}}{(2^{-k}+\rho(x, y_\tau^{k,\nu}))^{1+\epsilon}} + \frac{2^{-k\epsilon}}{(2^{-k}+\rho(x', y_\tau^{k,\nu}))^{1+\epsilon}}\Big],
\end{aligned}
$$

we obtain if $\rho(x,x') \le \frac{1}{2A}(2^{-k}+\rho(x,y))$ and $0<\epsilon'<\epsilon$,

$$\begin{aligned}
&|E_k(x,y)-E_k(x',y)| \\
&= \left|\sum_{\tau\in I_k}\sum_{\nu=1}^{N(k,\tau)}\int_{Q_\tau^{k,\nu}}[D_k^N(x,y_\tau^{k,\nu})-D_k^N(x',y_\tau^{k,\nu})]D_k(y_\tau^{k,\nu},y)d\mu(z)\right| \\
&\le \sum_{\tau\in I_k}\sum_{\nu=1}^{N(k,\tau)}\int_{Q_\tau^{k,\nu}}|D_k^N(x,y_\tau^{k,\nu})-D_k^N(x',y_\tau^{k,\nu})||D_k(y_\tau^{k,\nu},y)|d\mu(z) \\
&\le C\sum_{\tau\in I_k}\sum_{\nu=1}^{N(k,\tau)}\int_{Q_\tau^{k,\nu}}\Big(\frac{\rho(x,x')}{2^{-k}+\rho(x,y_\tau^{k,\nu})}\Big)^{\epsilon} \\
&\qquad\times\Big[\frac{2^{-k\epsilon}}{(2^{-k}+\rho(x,y_\tau^{k,\nu}))^{1+\epsilon}}+\frac{2^{-k\epsilon}}{(2^{-k}+\rho(x',y_\tau^{k,\nu}))^{1+\epsilon}}\Big] \\
&\qquad\times\frac{2^{-k\epsilon}}{(2^{-k}+\rho(y_\tau^{k,\nu},y))^{1+\epsilon}}d\mu(z) \\
&\le C\sum_{\tau\in I_k}\sum_{\nu=1}^{N(k,\tau)}\int_{Q_\tau^{k,\nu}}\left(\frac{\rho(x,x')}{2^{-k}+\rho(x,z)}\right)^{\epsilon} \\
&\qquad\times\Big[\frac{2^{-k\epsilon}}{(2^{-k}+\rho(x,z))^{1+\epsilon}}+\frac{2^{-k\epsilon}}{(2^{-k}+\rho(x',z))^{1+\epsilon}}\Big] \\
&\qquad\times\frac{2^{-k\epsilon}}{(2^{-k}+\rho(z,y))^{1+\epsilon}}d\mu(z) \\
&= C\int_X\Big(\frac{\rho(x,x')}{2^{-k}+\rho(x,z)}\Big)^{\epsilon}\Big[\frac{2^{-k\epsilon}}{(2^{-k}+\rho(x,z))^{1+\epsilon}}+\frac{2^{-k\epsilon}}{(2^{-k}+\rho(x',z))^{1+\epsilon}}\Big] \\
&\qquad\times\frac{2^{-k\epsilon}}{(2^{-k}+\rho(z,y))^{1+\epsilon}}d\mu(z) \\
&\le C\Big(\frac{\rho(x,x')}{2^{-k}+\rho(x,y)}\Big)^{\epsilon'}\frac{2^{-k\delta}}{(2^{-k}+\rho(x,y))^{1+\delta}}
\end{aligned}$$

where $\delta=\epsilon-\epsilon'>0$.

Similarly, if $\rho(y,y')\le\frac{1}{2A}(2^{-k}+\rho(x,y))$ and $0<\epsilon'<\epsilon$,

$$|E_k(x,y)-E_k(x,y')|\le C\Big(\frac{\rho(y,y')}{2^{-k}+\rho(x,y)}\Big)^{\epsilon'}\frac{2^{-k\delta}}{(2^{-k}+\rho(x,y))^{1+\delta}}$$

and, if $\rho(x,x')\le\frac{1}{2A}(2^{-k}+\rho(x,y)),\rho(y,y')\le\frac{1}{2A}(2^{-k}+\rho(x,y))$ and $0<\epsilon'<\epsilon$,

$$\begin{aligned}
&|E_k(x,y)-E_k(x',y)-E_k(x,y')+E_k(x',y')| \\
&\quad\le C\Big(\frac{\rho(x,x')}{2^{-k}+\rho(x,y)}\Big)^{\epsilon'}\Big(\frac{\rho(y,y')}{2^{-k}+\rho(x,y)}\Big)^{\epsilon'}\frac{2^{-k\delta}}{(2^{-k}+\rho(x,y))^{1+\delta}}.
\end{aligned}$$

It is easy to see that $E_k(1)=(E_k)^*(1)=0$. Let $T(f)=\sum_{|k|>M}E_k(f)$. Then T satisfies Theorem 2.4 with ϵ replaced by $\epsilon',0<\epsilon'<\epsilon$. As in the proof of

Theorem 3.15, it suffices to show that for $0 < \beta' < \beta$ and $0 < \gamma' < \gamma$ there exist a constant C which is independent of f and M, and some $\sigma > 0$ such that

$$\Big|\sum_{|k|>M} E_k(f)(x)\Big| \leq C2^{-\sigma M}(1+\rho(x,x_0))^{-(1+\gamma')}\|f\|_{\mathcal{M}(\beta,\gamma)}, \qquad (3.88)$$

and

$$\Big|\sum_{|k|>M} E_k(f)(x) - \sum_{|k|>M} E_k(f)(x')\Big| \qquad (3.89)$$
$$\leq C\Big(\frac{\rho(x,x')}{1+\rho(x,x_0)}\Big)^{\beta''}(1+\rho(x,x_0))^{-(1+\gamma')}\|f\|_{\mathcal{M}(\beta,\gamma)}$$

for $\rho(x,x') \leq \frac{1}{2}(1+\rho(x,x_0))$ and $\beta' < \beta'' < \epsilon'$.

Again, as in the proof of Theorem 3.15, (3.89) follows from Theorem 2.4 and (3.88) follows from the same proof for (3.75). We leave these details to the reader. The proof of Theorem 3.25 is complete.

By a duality argument, we obtain wavelet expansions on distribution space, which is given by the following theorem.

Theorem 3.26. *The wavelet expansions in Theorem 3.25 hold on distribution spaces $(\dot{\mathcal{M}}_0(\beta,\gamma))'$. More precisely, if $f \in (\dot{\mathcal{M}}_0(\beta',\gamma'))'$ then the wavelet expansions in (3.84) hold in $(\dot{\mathcal{M}}_0(\beta,\gamma))'$ with $\beta > \beta'$ and $\gamma > \gamma'$.*

As we mentioned for a continuous version of wavelet expansions in Theorem 3.15, the similar difficulties occur: one obstacle is that we cannot replace $L^2(X)$ by $L^\infty(X)$ in (3.84). However, these difficulties, as we have described before, will disappear once we establish wavelet expansions by starting with a scaling function. This variant of wavelet expansions is given by the following theorem.

Theorem 3.27. *There exist four families $\phi_\lambda, \widetilde{\phi}_\lambda, \bar{\phi}_\lambda, \bar{\bar{\phi}}_\lambda$ of scaling functions and six families $\psi_\lambda, \widetilde{\psi}_\lambda, \widetilde{\widetilde{\psi}}_\lambda, \bar{\psi}_\lambda, \bar{\bar{\psi}}_\lambda, \bar{\bar{\bar{\psi}}}_\lambda$ of wavelets such that for every square integrable function f one has*

$$\begin{aligned} f(x) &= \sum_{\lambda=(0,\tau,\nu)} \phi_\lambda(x)\langle f, \widetilde{\phi}_\lambda\rangle + \sum_{\lambda=(k,\tau,\nu),1\leq k\leq N} \phi_\lambda(x)\langle f, \widetilde{\psi}_\lambda\rangle \\ &\quad + \sum_{\lambda=(k,\tau,\nu),k\geq N+1} \widetilde{\widetilde{\psi}}_\lambda(x)\langle f, \psi_\lambda\rangle \\ &= \sum_{\lambda=(0,\tau,\nu)} \bar{\phi}_\lambda(x)\langle f, \bar{\bar{\phi}}_\lambda\rangle + \sum_{\lambda=(k,\tau,\nu),1\leq k\leq N} \bar{\phi}_\lambda(x)\langle f, \bar{\psi}_\lambda\rangle \\ &\quad + \sum_{\lambda=(k,\tau,\nu),k\geq N+1} \bar{\bar{\psi}}_\lambda(x)\langle f, \bar{\bar{\bar{\psi}}}_\lambda\rangle \end{aligned} \qquad (3.90)$$

where the series converge in the norm of $\mathcal{M}(\beta', \gamma')$ for $f \in \mathcal{M}(\beta, \gamma), 0 < \beta' < \beta$ and $0 < \gamma' < \gamma$, and the norm of $L^p(X)$, for $1 < p < \infty$.

Let $\{S_k\}_{k\in\mathbb{Z}}$ be an approximation to the identity and its kernels satisfy (3.19) with $\sigma = \epsilon \le \theta$. Set $D_k = S_k - S_{k-1}$ for $k \ge 1$, $D_0 = S_0$, and $D_k = 0$ for $k < 0$. As in the proof of Theorem 3.19, we have the following decomposition of the identity operator

$$\begin{aligned} I &= \sum_k D_k \sum_l D_l = \sum_k D_k^N D_k + \sum_k \sum_{|l|>N} D_{k+l} D_k \\ &= T_N + R_N, \end{aligned}$$

where $D_k^N = \sum\limits_{|l|\le N} D_{k+l}$.

Note that in Theorem 3.19, we have proved that

$$\|R_N(f)\|_{\mathcal{M}(\beta,\gamma)} \le C2^{-N\delta}\|f\|_{\mathcal{M}(\beta,\gamma)}.$$

Next, we decompose T_N as follows:

$$\begin{aligned} T_N(f)(x) &= \sum_k D_k^N D_k(f)(x) \\ &= \sum_{0\le k\le N} \int D_k^N(x,y) D_k(f)(y) d\mu(y) + \sum_{k>N} \int D_k^N(x,y) D_k(f)(y) d\mu(y) \\ &= \sum_{0\le k\le N} \sum_{\tau\in I_k} \sum_{v=1}^{N(k,\tau)} \mu(Q_\tau^{k,v}) m_{Q_\tau^{k,v}}(D_k^N(x,\cdot)) m_{Q_\tau^{k,v}}(D_k(f)) \\ &\quad + \sum_{k>N} \sum_{\tau\in I_k} \sum_{v=1}^{J(k,\tau)} \mu(Q_\tau^{k,v}) D_k^N(x, y_\tau^{k,v}) D_k(f)(y_\tau^{k,v}) + R_N^1(f)(x) \\ &\quad + R_N^2(f)(x) \end{aligned}$$

where $m_Q(f) = \frac{1}{\mu(Q)} \int\limits_Q f(x) d\mu(x)$,

$$\begin{aligned} R_N^1(f)(x) = \sum_{0\le k\le N} \sum_{\tau\in I_k} \sum_{v=1}^{N(k,\tau)} \int_{Q_\tau^{k,v}} &[D_k^N(x,y) D_k(f)(y) \\ &- m_{Q_\tau^{k,v}}(D_k^N(x,\cdot)) m_{Q_\tau^{k,v}}(D_k(f))] d\mu(y) \end{aligned}$$

and

$$\begin{aligned} R_N^2(f)(x) = \sum_{k>N} \sum_{\tau\in I_k} \sum_{v=1}^{J(k,\tau)} \int_{Q_\tau^{k,v}} &[D_k^N(x,y) D_k(f)(y) \\ &- D_k^N(x, y_\tau^{k,v}) D_k(f)(y_\tau^{k,v})] d\mu(y). \end{aligned}$$

By the same proof as in Lemma 3.23, $R_N^2(x,y)$, the kernel of R_N^2, satisfies (3.70)-(3.74), and $R_N^2(1) = 0$, $(R_N^2)^*(1) = 0$, because $\int D_k^N(x,y)d\mu(x) = \int D_k^N(x,y)d\mu(y) = 0$ for $k > N$. Hence $\|R_N^2(f)\|_{\mathcal{M}(\beta,\gamma)} \leq C_N 2^{-j\delta}\|f\|_{\mathcal{M}(\beta,\gamma)}$, where C_N is a constant depending on N.

Now we estimate R_N^1. To this end, we write

$$\begin{aligned} &R_N^1(f)(x)\\ &= \sum_{0\leq k\leq N}\sum_{\tau\in I_k}\sum_{v=1}^{N(k,\tau)}\int_{Q_\tau^{k,v}}[D_k^N(x,y) - m_{Q_\tau^{k,v}}(D_k^N(x,\cdot))]D_k(f)(y)d\mu(y)\\ &\quad+\sum_{0\leq k\leq N}\sum_{\tau\in I_k}\sum_{v=1}^{N(k,\tau)}\int_{Q_\tau^{k,v}} m_{Q_\tau^{k,\nu}}(D_k^N(x,\cdot))[D_k(f)(y) - m_{Q_\tau^{k,v}}(D_k(f))]d\mu(y)\\ &= R_N^{1,1}(f)(x) + R_N^{1,2}(f)(x). \end{aligned}$$

Since the estimates for $R_N^{1,1}$ and $R_N^{1,2}$ are similar, we only estimate $R_N^{1,2}$.

It is easy to check that if the kernel of $R_N^{1,2}$ is denoted by $R_N^{1,2}(x,y)$, then

$$\begin{aligned} R_N^{1,2}(x,y) = \sum_{0\leq k\leq N}\sum_{\tau\in I_k}\sum_{v=1}^{N(k,\tau)}\frac{1}{\mu(Q_\tau^{k,v})}\int_{Q_\tau^{k,v}}\int_{Q_\tau^{k,v}} D_k^N(x,z)\\ \times[D_k(w,y) - D_k(z,y)]d\mu(w)d\mu(z). \end{aligned}$$

By the same proof as in Lemma 3.23, $R_N^{1,2}(x,y)$ satisfies (3.70)-(3.74). All we need to check is that $R_N^{1,2}(1) = (R_N^{1,2})^*(1) = 0$. In fact, we have

$$\begin{aligned} R_N^{1,2}(1)(x) = \sum_{0\leq k\leq N}\sum_{\tau\in I_k}\sum_{v=1}^{N(k,\tau)}\int_{Q_\tau^{k,v}} m_{Q_\tau^{k,\nu}}(D_k^N(x,\cdot))\\ \times[D_k(1)(y) - m_{Q_\tau^{k,v}}(D_k(1))]d\mu(y). \end{aligned}$$

Note that $D_k(1)(y) = \int D_k(z,y)d\mu(z) = 0$ for $1 \leq k \leq N$ and that $\int D_0(z,y)d\mu(z) = \int S_0(z,y)d\mu(z) = 1$, this implies $[D_k(1)(y) - m_{Q_\tau^{k,v}}(D_k(1))] = 0$ for $0 \leq k \leq N$. Hence $R_N^{1,2}(1) = 0$.

As to $(R_N^{1,2})^*(1)$, note that $\int D_k^N(x,z)d\mu(x) = 1$ for $0 \leq k \leq N$, so

$$\begin{aligned} &(R_N^{1,2})^*(1)(y)\\ &= \sum_{0\leq k\leq N}\sum_{\tau\in I_k}\sum_{v=1}^{N(k,\tau)}\frac{1}{\mu(Q_\tau^{k,v})}\int_{Q_\tau^{k,v}}\int_{Q_\tau^{k,v}}[D_k(w,y) - D_k(z,y)]d\mu(w)d\mu(z) = 0. \end{aligned}$$

By the same proof as R_N^2, we obtain $\|R_N^{1,2}(f)\|_{\mathcal{M}(\beta,\gamma)} \leq C_N 2^{-j\delta}\|f\|_{\mathcal{M}(\beta,\gamma)}$, where C_N is a constant depending on N. Similarly, as we mentioned, $\|R_N^{1,1}(f)\|_{\mathcal{M}(\beta,\gamma)} \leq C_N 2^{-j\delta}\|f\|_{\mathcal{M}(\beta,\gamma)}$.

We now define the operator S by

$$S(f)(x) = \sum_{0\le k\le N}\sum_{\tau\in I_k}\sum_{v=1}^{N(k,\tau)} \mu(Q_\tau^{k,v}) m_{Q_\tau^{k,v}}(D_k^N(x,\cdot)) m_{Q_\tau^{k,v}}(D_k(f)) + \sum_{k>N}\sum_{\tau\in I_k}\sum_{v=1}^{J(k,\tau)} \mu(Q_\tau^{k,v}) D_k^N(x, y_\tau^{k,v}) D_k(f)(y_\tau^{k,v}).$$

Let $R(f)(x) = R_N(f) + R_N^1(f)(x) + R_N^2(f)(x)$. Then we get $I = S + R$. Now first choose N sufficiently large, and then choose j such that $\|R\|_{\mathcal{M}(\beta,\gamma)\to\mathcal{M}(\beta,\gamma)} < 1$. Hence S is invertible on $\mathcal{M}(\beta,\gamma)$ and $S^{-1} = \sum_{m=0}^{\infty} R^m$.

Define $\phi_\lambda(x) = m_{Q_\tau^{k,v}} S^{-1}(D_k^N(\cdot,\cdot))(x)$ for $0 \le k \le N, \widetilde{\phi}_\lambda(x) = m_{Q_\tau^{0,v}}(S_0(\cdot,x)), \widetilde{\psi}_\lambda(x) = m_{Q_\tau^{k,v}}(D_k(\cdot,x))$ for $1 \le k \le N$, and $\widetilde{\widetilde{\psi}}_\lambda(x) = S^{-1}(D_k^N(\cdot, y_\tau^{k,v})(x), \psi_\lambda = D_k(y_\tau^{k,v}, x)$ for $k \ge N+1$. We obtain the first identity in Theorem 3.27.

To see that the series converge in the norm of $\mathcal{M}(\beta',\gamma')$ for $0 < \beta' < \beta$ and $0 < \gamma' < \gamma$, and in the spaces of $L^p(X)$, for $1 < p < \infty$, we write

$$\begin{aligned}
&\sum_{0\le k\le N}\sum_{\tau\in I_k}\sum_{v=1}^{N(k,\tau)} \mu(Q_\tau^{k,v}) m_{Q_\tau^{k,v}}(\tilde{S}_k(x,\cdot)) m_{Q_\tau^{k,v}}(D_k(f)) \\
&\quad + \sum_{k=N+1}^{L}\sum_{\tau\in I_k}\sum_{v=1}^{J(k,\tau)} \mu(Q_\tau^{k,v}) \tilde{D}_k(x, y_\tau^{k,v}) D_k(f)(y_\tau^{k,v}) \\
&= S^{-1}\Bigg(\sum_{0\le k\le N}\sum_{\tau\in I_k}\sum_{v=1}^{N(k,\tau)} \mu(Q_\tau^{k,v}) m_{Q_\tau^{k,v}}(D_k^N(x,\cdot)) m_{Q_\tau^{k,v}}(D_k(f)) \\
&\quad + \sum_{k=N+1}^{L}\sum_{\tau\in I_k}\sum_{v=1}^{J(k,\tau)} \mu(Q_\tau^{k,v}) D_k^N(x, y_\tau^{k,v}) D_k(f)(y_\tau^{k,v})\Bigg) \\
&= S^{-1}\Bigg(S(f) - \sum_{k=L+1}^{\infty}\sum_{\tau\in I_k}\sum_{v=1}^{J(k,\tau)} \mu(Q_\tau^{k,v}) D_k^N(x, y_\tau^{k,v}) D_k(f)(y_\tau^{k,v})\Bigg).
\end{aligned}$$

Then the conclusion follows from the same proof of Theorem 3.25. The proof for the second identity is the same. We leave the details to reader.

Chapter 4
Wavelets and Spaces of Functions and Distributions

4.1 Introduction

In Chapter 3 we built wavelet expansions on a space of homogeneous type, which was one of main goals in this book. These wavelets are not orthonormal bases, but wavelet frames $\psi_\lambda, \lambda \in \Lambda$. These wavelets are (i) localized, (ii) smooth and (iii) oscillating. These oscillations are described by the fundamental cancellation property (iv) $\int\limits_X \psi_\lambda(x)d\mu(x) = 0$. This being said, the expansion of a function $f \in L^2(X, d\mu)$ into a wavelet series is given by

$$f(x) = \sum_{\lambda\in\Lambda} a(\lambda)\widetilde{\psi}_\lambda(x) \tag{4.1}$$

where

$$\|f\|_2 \approx \left\{ \sum_{\lambda\in\Lambda} |a(\lambda)|^2 \right\}^{\frac{1}{2}} \tag{4.2}$$

and where the coefficients $a(\lambda)$ are given by

$$a(\lambda) = \int f(x)\psi_\lambda(x)d\mu(x). \tag{4.3}$$

The key point is the following. The dual wavelets $\widetilde{\psi}_\lambda$ are sharing with ψ_λ the same localization, smoothness and vanishing integral properties. This implies that most functional spaces will be characterized by simple size properties of the wavelet coefficients in (4.3). These wavelet expansions and characterization of functional spaces will constitute the heart of this book, which will be given in this chapter.

In order to define or characterize some spaces of functions and distributions by size properties of wavelet coefficients, one should first prove that

D. Deng and Y. Han, *Harmonic Analysis on Spaces of Homogeneous Type*, Lecture Notes in Mathematics 1966,

these size properties do not depend on the choice of wavelets. We will first prove a comparison theorem for these size properties of the wavelet coefficients. In the next four Sections, 4.3-4.6, we will use the size properties of wavelet expansions to study the classical Hölder spaces, Lebesgue space L^p, generalized Sobolev spaces, Hardy spaces H^p, BMO and Besov spaces on spaces of homogeneous type. As we mentioned, wavelet expansions provide a bridge between operators and spaces of functions and distributions. In Section 4.7, we will prove the $T1$ type theorems, that is, the boundedness of Calderón-Zygmund operators on these spaces. As we described in Chapter 3, the condition (3.19), namely the double Lipschitz condition, is crucial for the construction of wavelet expansions. This condition, however, is not needed for establishing the Littlewood-Paley estimates obtained by G. David, J. L. Journé and S. Semmes, on $L^p(X, d\mu), 1 < p < \infty$. A natural question arises: Can these spaces be characterized without using the condition (3.19)? As a beautiful application of the $T1$ type theorems, we will give a positive answer for this question in the last Section 4.7.

4.2 Comparison Properties of Wavelet Coefficients

Suppose that both ψ_λ, and $\overline{\psi}_\lambda, \lambda \in \Lambda$, are wavelets as given in Theorem 3.25. $\langle f, \psi_\lambda \rangle$ and $\langle f, \overline{\psi}_\lambda \rangle$ are the wavelet coefficients of f with respect to wavelets ψ_λ and $\overline{\psi}_\lambda$, respectively. We prove the following comparisons of the size properties of the wavelet coefficients. See [H2] for more details.

Theorem 4.1 (the comparison property). *Suppose $f \in \big(\dot{\mathcal{M}}_0(\beta, \gamma)\big)'$ with $0 < \beta, \gamma < \theta$. Then for $-\theta < s < \theta, \max\left(\frac{1}{1+\theta}, \frac{1}{1+\theta+s}\right) < p < \infty$,*

$$\left\| \left\{ \sum_{\lambda \in \Lambda} \left[2^{k(s+\frac{1}{2})} |\langle f, \psi_\lambda \rangle| \chi_{Q_\lambda} \right]^2 \right\}^{\frac{1}{2}} \right\|_p \approx \left\| \left\{ \sum_{\lambda \in \Lambda} \left[2^{k(s+\frac{1}{2})} |\langle f, \overline{\psi}_\lambda \rangle| \chi_{Q_\lambda} \right]^2 \right\}^{\frac{1}{2}} \right\|_p \tag{4.4}$$

and if $-\theta < s < \theta, \max\left(\frac{1}{1+\theta}, \frac{1}{1+\theta+s}\right) < p \le \infty$ and $0 < q \le \infty$,

$$\left\{ \sum_{k \in \mathbb{Z}} \left(\sum_{\lambda \in \Lambda_k} \left[2^{k(s+\frac{1}{2}-\frac{1}{p})} |\langle f, \psi_\lambda \rangle| \right]^p \right)^{\frac{q}{p}} \right\}^{\frac{1}{q}} \approx \left\{ \sum_{k \in \mathbb{Z}} \left(\sum_{\lambda \in \Lambda_k} \left[2^{k(s+\frac{1}{2}-\frac{1}{p})} |\langle f, \overline{\psi}_\lambda \rangle| \right]^p \right)^{\frac{q}{p}} \right\}^{\frac{1}{q}} \tag{4.5}$$

where $\Lambda = \{(k, \tau, \nu) : k \in \mathbb{Z}, \tau \in I_k, 1 \le \nu \le N(k, \tau)\}$ and $\Lambda_k = \{(k, \tau, \nu) : \tau \in I_k, 1 \le \nu \le N(k, \tau)\}$.

The key idea to prove Theorem 4.1 is to use the almost orthogonal argument for wavelets and wavelet expansions. We first prove the comparison (4.4). Using the wavelet expansion given by wavelets $\overline{\psi}_\lambda$, we have

$$\langle f, \psi_\lambda \rangle = \sum_{\lambda'} \langle \widetilde{\overline{\psi}}_{\lambda'}, \psi_\lambda \rangle \langle f, \overline{\psi}_{\lambda'} \rangle.$$

By the almost orthogonal argument in Lemma 3.7, there exists a constant C such that

$$|\langle \widetilde{\overline{\psi}}_{\lambda'}, \psi_\lambda \rangle| \leq C 2^{-\frac{k}{2}} 2^{-\frac{k'}{2}} 2^{-|k-k'|\epsilon} \frac{2^{-(k\wedge k')\epsilon}}{(2^{-(k\wedge k')} + \rho(y_\lambda, y_{\lambda'}))^{1+\epsilon}} \tag{4.6}$$

where $|s| < \epsilon < \theta, \max(\frac{1}{1+\epsilon}, \frac{1}{1+\epsilon+s}) < p < \infty, \lambda \in \Lambda_k, \lambda' \in \Lambda_{k'}, y_\lambda$, and $y_{\lambda'}$ are any points which, as in Theorem 3.25, are chosen from Q_λ and $Q_{\lambda'}$, respectively. Thus,

$$\begin{aligned} &|\langle f, \psi_\lambda \rangle| \\ &\leq C \sum_{k' \in \mathbb{Z}} \sum_{\lambda' \in \Lambda_{k'}} 2^{-\frac{k}{2}} 2^{-\frac{k'}{2}} 2^{-|k-k'|\epsilon} \frac{2^{-(k\wedge k')\epsilon}}{(2^{-(k\wedge k')} + \rho(y_\lambda, y_{\lambda'}))^{1+\epsilon}} |\langle f, \overline{\psi}_{\lambda'} \rangle|. \end{aligned} \tag{4.7}$$

We get

$$\begin{aligned} &\left\{ \sum_{\lambda \in \Lambda} \left[2^{k(s+\frac{1}{2})} |\langle f, \psi_\lambda \rangle| \chi_{Q_\lambda} \right]^2 \right\}^{\frac{1}{2}} \\ &\quad \leq C \left\{ \sum_{\lambda \in \Lambda} \left[\sum_{k' \in \mathbb{Z}} \sum_{\lambda' \in \Lambda_{k'}} 2^{ks} 2^{-|k-k'|\epsilon - \frac{k'}{2}} \right.\right. \\ &\qquad \left.\left. \cdot \frac{2^{-(k\wedge k')\epsilon}}{(2^{-(k\wedge k')} + \rho(x, y_{\tau'}^{k',\nu'}))^{1+\epsilon}} |\langle f, \overline{\psi}_{\lambda'} \rangle| \chi_{Q_\lambda} \right]^2 \right\}^{\frac{1}{2}}. \end{aligned}$$

By an estimate in [FJ],

$$\begin{aligned} &\sum_{\tau' \in I_{k'}} \sum_{\nu'=1}^{N(k',\tau')} \frac{2^{-(k\wedge k')\epsilon}}{(2^{-(k\wedge k')} + \rho(x, y_{\tau'}^{k',\nu'}))^{1+\epsilon}} |\langle f, \overline{\psi}_{\lambda'} \rangle| \\ &\quad \leq C 2^{(k\wedge k') + \frac{[k' - (k\wedge k')]}{r}} \left\{ M \left(\sum_{\tau' \in I_{k'}} \sum_{\nu'=1}^{N(k',\tau')} |\langle f, \overline{\psi}_{\lambda'} \rangle| \chi_{Q_{\lambda'}} \right)^r \right\}^{\frac{1}{r}} (x) \end{aligned}$$

where $p > r > \frac{1}{1+\epsilon}$.

Putting this estimate into the last term above and applying Hölder's inequality yield

$$\left\{\sum_{\lambda\in\Lambda}\left[2^{k(s+\frac{1}{2})}|\langle f,\psi_\lambda\rangle|\chi_{Q_\lambda}\right]^2\right\}^{\frac{1}{2}}$$
$$\leq C\left\{\sum_{k\in\mathbb{Z}}\left[\sum_{k'\in\mathbb{Z}}2^{(k-k')s-|k-k'|\epsilon}2^{\frac{(k\wedge k')+[k'-(k\wedge k')]}{r}}\right.\right.$$
$$\left.\left.\times\left[M\left(\sum_{\lambda'\in\Lambda_{k'}}2^{k'(s+\frac{1}{2})}|\langle f,\overline{\psi}_{\lambda'}\rangle|\ \chi_{Q_{\lambda'}}\right)^r\right]^{\frac{1}{r}}(x)\right]^2\right\}^{\frac{1}{2}}$$
$$\leq C\left\{\sum_{k'\in\mathbb{Z}}\left[M\left(\sum_{\lambda'\in\Lambda_{k'}}2^{k'(s+\frac{1}{2})}|\langle f,\overline{\psi}_{\lambda'}\rangle|\chi_{Q_{\lambda'}}\right)^r\right]^{\frac{2}{r}}(x)\right\}^{\frac{1}{2}}$$

where we use the facts that $|s|<\epsilon, \max\{\frac{1}{1+\epsilon},\frac{1}{1+\epsilon+s}\}<r<p$, and

$$\sup_k\sum_{k'}2^{(k-k')s-|k-k'|\epsilon+(k\wedge k')+\frac{[k'-(k\wedge k')]}{r}}<\infty,$$
$$\sup_{k'}\sum_{k}2^{(k-k')s-|k-k'|\epsilon+(k\wedge k')+\frac{[k'-(k\wedge k')]}{r}}<\infty.$$

Now, the Fefferman-Stein vector-valued maximal function inequality [FS] implies

$$\left\|\left\{\sum_{\lambda\in\Lambda}\left[2^{k(s+\frac{1}{2})}|\langle f,\psi_\lambda\rangle|\chi_{Q_\tau^{k,\nu}}\right]^2\right\}^{\frac{1}{2}}\right\|_p$$
$$\leq C\left\|\left\{\sum_{\lambda\in\Lambda}\left[2^{k(s+\frac{1}{2})}|\langle f,\overline{\psi}_\lambda\rangle|\chi_{Q_\tau^{k,\nu}}\right]^2\right\}^{\frac{1}{2}}\right\|_p$$

which implies one implication in (4.4). The other implication in (4.4) follows from the same proof.

We now prove (4.5). If $p>\frac{1}{1+\epsilon}$, using the estimates in (4.6) and (4.7) and applying Hölder's inequality for $p>1$ and the triangle inequality $(a+b)^p\leq a^p+b^p$ for $p\leq 1$, we have

$$\sum_{\tau'\in I_{k'}}\sum_{\nu'=1}^{N(k',\tau')}\mu(Q_{\tau'}^{k',\nu'})\left[\frac{2^{-(k\wedge k')\epsilon}}{(2^{-(k\wedge k')}+\rho(z_\tau^{k,\nu},y_{\tau'}^{k',\nu'}))^{1+\epsilon}}\right]^{(p\wedge 1)}$$
$$\leq C\int\left[\frac{2^{-(k\wedge k')\epsilon}}{(2^{-(k\wedge k')}+\rho(z_\tau^{k,\nu},y))^{1+\epsilon}}\right]^{(p\wedge 1)}d\mu(y)$$
$$\leq C2^{-(k\wedge k')[1-((p\wedge 1)]}.$$

Therefore, by the estimates in (4.6) and (4.7), we obtain

$$\sum_{\lambda\in\Lambda_k}\left[2^{k(s+\frac{1}{2}-\frac{1}{p})}|\langle f,\psi_\lambda\rangle|\right]^p$$
$$\leq C\sum_{\lambda\in\Lambda_k}\left[\sum_{k'\in\mathbb{Z}}\sum_{\lambda'\in\Lambda_{k'}}2^{(k-k')s-|k-k'|\epsilon-\frac{k}{p}-k'}\right.$$
$$\left.\times\frac{2^{-(k\wedge k')\epsilon}}{(2^{-(k\wedge k')}+\rho(y_\lambda,y_{\lambda'}))^{1+\epsilon}}2^{k'(s+\frac{1}{2}}|\langle f,\overline{\psi}_{\lambda'}\rangle|\right]^p$$
$$\leq C\sum_{k'\in\mathbb{Z}}\sum_{\lambda'\in\Lambda_{k'}}2^{(k-k')s(p\wedge1)-|k-k'|\epsilon(p\wedge1)-(k\wedge k')[1-(p\wedge1)]-k'(p\wedge1)}$$
$$\times[2^{k'(s+\frac{1}{2})}|\langle f,\overline{\psi}_{\lambda'}\rangle|]^p.$$

Finally, by Hölder's inequality again for $q/p>1$ and the triangle inequality $(a+b)^{q/p}\leq a^{q/p}+b^{q/p}$ for $q/p\leq1$, we get

$$\left\{\sum_{k\in\mathbb{Z}}\left(\sum_{\lambda\in\Lambda_k}\left[2^{k'(s+\frac{1}{2}-\frac{1}{p})}|\langle f,\psi_\lambda\rangle|\right]^p\right)^{\frac{q}{p}}\right\}^{\frac{1}{q}}$$
$$\leq C\left\{\sum_{k\in\mathbb{Z}}\left(\sum_{k'\in Z}2^{(k-k')s(p\wedge1)-|k-k'|\epsilon(p\wedge1)-(k\wedge k')[1-(p\wedge1)]-k'(p\wedge1)}\right.\right.$$
$$\left.\left.\times\sum_{\lambda'\in\Lambda_{k'}}[2^{k'(s+\frac{1}{2}-\frac{1}{p})}|\langle f,\overline{\psi}_{\lambda'}\rangle|]^p\right)^{\frac{q}{p}}\right\}^{\frac{1}{q}}$$
$$\leq C\left\{\sum_{k\in\mathbb{Z}}\sum_{k'\in Z}(2^{(k-k')s(p\wedge1)-|k-k'|\epsilon(p\wedge1)-(k\wedge k')[1-(p\wedge1)]-k'(p\wedge1)})^{(\frac{q}{p}\wedge1)}\right.$$
$$\left.\times\left(\sum_{\lambda'\in\Lambda_{k'}}[2^{k'(s+\frac{1}{2}-\frac{1}{p})}|\langle f,\overline{\psi}_{\lambda'}\rangle|]^p\right)^{\frac{q}{p}}\right\}^{\frac{1}{q}}$$
$$\leq C\left\{\sum_{k'\in\mathbb{Z}}\left(\sum_{\lambda'\in\Lambda_{k'}}[2^{k'(s+\frac{1}{2}-\frac{1}{p})}|\langle f,\overline{\psi}_{\lambda'}\rangle|]^p\right)^{\frac{q}{p}}\right\}^{\frac{1}{q}}$$

where we use the facts that if $|s|<\epsilon$ and $p>\frac{1}{1+s+\epsilon}$,

$$\sup_k\sum_{k'}2^{(k-k')s(p\wedge1)-|k-k'|\epsilon(p\wedge1)-(k\wedge k')[1-(p\wedge1)]-k'(p\wedge1)}<\infty$$

and

$$\sup_{k'}\sum_{k}(2^{(k-k')s(p\wedge1)-|k-k'|\epsilon(p\wedge1)-(k\wedge k')[1-(p\wedge1)]-k'(p\wedge1)})^{(\frac{q}{p}\wedge1)}<\infty.$$

This implies the proof of one implication in (4.5). The proof of the other implication in (4.5) is similar. The proof of Theorem 4.1 is complete.

The key fact used in the proof of Theorem 4.1 is the wavelet expansion provided by Theorem 3.25. As we already noticed, this expansion does not converge in some functional spaces, like L^1 or L^∞. This can be improved by using the wavelet expansion described in Theorem 3.27. This means that we will prove the following "inhomogeneous comparison" of the size properties of wavelet coefficients.

Suppose that $\phi_\lambda, \widetilde{\phi}_\lambda, \overline{\phi}_\lambda, \widetilde{\overline{\phi}}_\lambda$ are scaling functions and $\psi_\lambda, \widetilde{\psi}_\lambda, \widetilde{\widetilde{\psi}}_\lambda, \overline{\psi}_\lambda, \widetilde{\overline{\psi}}_\lambda, \widetilde{\widetilde{\overline{\psi}}}_\lambda$ are wavelets such that Theorem 3.27 holds:

$$\begin{aligned} f(x) &= \sum_{\lambda=(0,\tau,\nu)} \phi_\lambda(x)\langle f, \widetilde{\phi}_\lambda\rangle + \sum_{\lambda=(k,\tau,\nu),1\le k\le N} \phi_\lambda(x)\langle f, \widetilde{\psi}_\lambda\rangle \qquad (4.8)\\ &\quad + \sum_{\lambda=(0,\tau,\nu),k\ge N+1} \widetilde{\widetilde{\psi}}_\lambda(x)\langle f, \psi_\lambda\rangle \\ &= \sum_{\lambda=(0,\tau,\nu)} \overline{\phi}_\lambda(x)\langle f, \widetilde{\overline{\phi}}_\lambda\rangle + \sum_{\lambda=(k,\tau,\nu),1\le k\le N} \overline{\phi}_\lambda(x)\langle f, \widetilde{\overline{\psi}}_\lambda\rangle \\ &\quad + \sum_{\lambda=(0,\tau,\nu),k\ge N+1} \widetilde{\widetilde{\overline{\psi}}}_\lambda(x)\langle f, \overline{\psi}_\lambda\rangle. \end{aligned}$$

The inhomogeneous comparison of the size properties of the above two wavelet expansions is given by the following theorem.

Theorem 4.2 (the inhomogeneous comparison). *Suppose that* $f \in (\dot{\mathcal{M}}(\beta,\gamma))'$ *with* $0<\beta,\gamma<\theta$, $-\theta<s<\theta$. *Then if* $\max\left(\frac{1}{1+\theta}, \frac{1}{1+\theta+s}\right) < p < \infty$ *and* $\max\left(\frac{1}{1+\theta}, \frac{1}{1+\theta+s}\right) < q < \infty$,

$$\begin{aligned} &\Big\{\sum_{\lambda=(0,\tau,\nu)} |\langle f, \widetilde{\phi}_\lambda|^p\Big\}^{1/p} + \Big\|\Big\{\sum_{\lambda=(k,\tau,\nu),1\le k\le N} \big[2^{k(s+\frac12)}|\langle f, \widetilde{\psi}_\lambda|\chi_{Q_\lambda}(\cdot)\big]^q\Big\}^{1/q}\Big\|_p \\ &\qquad + \Big\|\Big\{\sum_{\lambda=(k,\tau,\nu),k\ge N+1} \big[2^{k(s+\frac12)}|\langle f, \psi_\lambda|\chi_{Q_\lambda}(\cdot)\big]^q\Big\}^{1/q}\Big\|_p \\ &\approx \Big\{\sum_{\lambda=(0,\tau,\nu)} |\langle f, \widetilde{\overline{\phi}}_\lambda|^p\Big\}^{1/p} + \Big\|\Big\{\sum_{\lambda=(k,\tau,\nu),1\le k\le N} \big[2^{k(s+\frac12)}|\langle f, \widetilde{\overline{\psi}}_\lambda|\chi_{Q_\lambda}(\cdot)\big]^q\Big\}^{1/q}\Big\|_p \\ &\qquad + \Big\|\Big\{\sum_{\lambda=(k,\tau,\nu),k\ge N+1} \big[2^{k(s+\frac12)}|\langle f, \overline{\psi}_\lambda|\chi_{Q_\lambda}(\cdot)\big]^q\Big\}^{1/q}\Big\|_p \end{aligned}$$

and if $\max\left(\frac{1}{1+\theta}, \frac{1}{1+\theta+s}\right) < p \le \infty$ *and* $\max\left(\frac{1}{1+\theta}, \frac{1}{1+\theta+s}\right) < q \le \infty$, *then*

$$\left\{\sum_{\lambda=(0,\tau,\nu)} |\langle f, \widetilde{\phi}_\lambda|^p\right\}^{1/p} + \left\{\sum_{k=1}^{N}\sum_{\tau\in I_k}\sum_{\nu=1}^{N(k,\tau)} \left[2^{k(s+\frac{1}{2-\frac{1}{p}})}|\langle f, \widetilde{\psi}_\lambda|\right]^p\right\}^{q/p}\Bigg\}^{1/p}$$

$$+\left\{\sum_{k=N+1}^{\infty}\sum_{\tau\in I_k}\sum_{\nu=1}^{N(k,\tau)} \left[2^{k(s+\frac{1}{2-\frac{1}{p}})}|\langle f, \psi_\lambda|\right]^p\right\}^{q/p}\Bigg\}^{1/p}$$

$$\approx\left\{\sum_{\lambda=(0,\tau,\nu)} |\langle f, \widetilde{\widetilde{\phi}}_\lambda|^p\right\}^{1/p} + \left\{\sum_{k=1}^{N}\sum_{\tau\in I_k}\sum_{\nu=1}^{N(k,\tau)} \left[2^{k(s+\frac{1}{2-\frac{1}{p}})}|\langle f, \widetilde{\widetilde{\psi}}_\lambda|\right]^p\right\}^{q/p}\Bigg\}^{1/p}$$

$$+\left\{\sum_{k=N+1}^{\infty}\sum_{\tau\in I_k}\sum_{\nu=1}^{N(k,\tau)} \left[2^{k(s+\frac{1}{2-\frac{1}{p}})}|\langle f, \overline{\psi}_\lambda|\right]^p\right\}^{q/p}\Bigg\}^{1/p}.$$

The proof of Theorem 4.2 is similar to the proof of Theorem 4.1. One instead uses the wavelet expansions in (4.8). The almost orthogonal argument is not needed when we deal with terms with $0 \le k \le N$. This almost orthogonal argument is only used to deal with the last terms where $N+1 \le k < \infty$. We leave these details to the reader.

In the following sections, based on Theorem 4.1 and Theorem 4.2, we will use fixed wavelet expansions. As in the standard case of $\mathbb{R}^n$, the advantage of wavelet expansions is the ease with which they adapt to various commonly used functional analysis norms. That is, if f belongs to one of the classical spaces, such as Hölder spaces, Sobolev spaces, Hardy spaces, or Besov spaces, the corresponding wavelet expansions will converge automatically to f in the appropriate norm. These properties will be described in the next sections.

4.3 Hölder Spaces

We will follow [M2] in this section. Let $0 < s < \theta$ where θ is the regularity of a quasi-metric ρ given in (1.7). There are two kinds of Hölder spaces, that is, the homogeneous Hölder spaces $\dot{C}^s(X)$ and the inhomogeneous Hölder spaces $C^s(X)$. We first recall the definition of the homogeneous Hölder spaces $\dot{C}^s(X), 0 < s < \theta$. For a continuous function $f : X \to \mathbb{C}$, we define the modulus of continuity $\omega_f(h)$ by $\omega_f(h) = \sup\limits_{\rho(x,y)\le h} |f(x) - f(y)|$. Then $f \in \dot{C}^s(X)$ if and only if there exists a constant C such that, for every $h > 0$, we have $\omega_f(h) \le Ch^s$. The norm of f in $\dot{C}^s(X)$ is defined by the lower bound of the constants C. Note that the norm we just defined is not a norm, because the constant functions have norm zero. This means that the homogeneous Hölder space $\dot{C}^s(X)$ is not a functional space, instead it is a space of equivalent classes modulo the constant functions. Finally, we define $\dot{C}^s(X), 0 < s < \theta$, to be the quotient space of the above function space modulo the constant functions.

To prove that such a defined $\dot{C}^s(X)$ is a Banach space, we have to define an operator, that is, for each equivalent class there is a corresponding function f_0 such that it is a representative of the class f and satisfies the inequalities $\sup_K |f_0| \leq C(K)\|f\|_{\dot{C}^s}$, where K is an arbitrary compact subset in X and the constant $C(K)$ depends only on K. These inequalities provide the following proof of the complement of $\dot{C}^s(X)$. Given a series $\sum_{j=1}^{\infty} f_j$, with $\|f_j\|_{\dot{C}^s} \leq C2^{-j}$. It suffices to show that the series converges to function $f \in \dot{C}^s$. To do this, for each class of f_j we can choose a representative $f_{j,0}$ such that $\sup_K |f_{j,0}| \leq C(K)2^{-j}$, which implies that the series $\sum_1^{\infty} f_{j,0}$ converges uniformly to a function f_0 on K. It is easy to see that $f_0 \in \dot{C}^s$. Therefore f_0 is a representative of an equivalence class of $\dot{C}^s$, and hence $\dot{C}^s$ is a Banach space.

The simplest way to choose a representative of a class $f \in \dot{C}^s$ is to fix any point $x_0 \in X$ and then choose a function f_0 in a class in $\dot{C}^s$ so that $f_0(x_0) = 0$, where x_0 is any fixed point in X. This can be done by setting $f_0(x) = f(x) - f(x_0)$ for $f \in \dot{C}^s$. Thus, for all $x \in X, |f_0(x)| \leq C(\rho(x, x_0))^s$. From this, the inequality $\sup_K |f_0| \leq C(K)\|f\|_{\dot{C}^s}$ follows easily.

Before giving a characterization of $\dot{C}^s$ in terms of a wavelet expansion, we come to the Littlewood-Paley analysis developed by G. David, J. L. Journé and S. Semmes. The following lemma gives a characterization of $\dot{C}^s$.

Lemma 4.3. *Let $0 < s < \theta$. Suppose that $\{D_k(x, y)\}_{k\in\mathbb{Z}}$ with the regularity order θ, is a family of mother functions given in Theorem 3.18. Let $f \in (\dot{\mathcal{M}}_0(\beta, \gamma))', s < \beta, \gamma < \theta$ such that by Theorem 3.18, $f = \sum_k \widetilde{D}_k D_k(f)$, in the sense of converges in $(\dot{\mathcal{M}}_0(\beta', \gamma'))', \beta < \beta' < \theta, \gamma < \gamma' < \theta$. Then f belongs to $\dot{C}^s, 0 < s < \theta$, if and only if $\|D_k(f)\|_\infty \leq C2^{-ks}$, for some constant C and for all $k \in \mathbb{Z}$.*

Suppose first that $f \in \dot{C}^s, 0 < s < \theta$. We may assume that $f(x_0) = 0$ for some fixed point $x_0 \in X$. The above proof of the complement of $\dot{C}^s$ tells us that $f(x)$ grows slowly at infinity. Therefore f is a distribution in $(\dot{\mathcal{M}}_0(\beta, \gamma))', s < \beta, \gamma < \theta$. We have

$$\begin{aligned} D_k(f)(x) &= \int D_k(x, y)f(y)d\mu(y) \\ &= \int D_k(x, y)[f(y) - f(x)]d\mu(y), \end{aligned}$$

because $D_k(x, y)$ is a wavelet and its integral is zero. Thus

$$\|D_k(f)\|_\infty \le C \int |D_k(x,y)|\rho^s(x,y)d\mu(y) \le C2^{-ks}.$$

We now prove the converse implication of Lemma 4.3. Suppose that $f = \sum_k \widetilde{D}_k D_k(f)$, in the sense of converges in $(\dot{\mathcal{M}}_0(\beta',\gamma'))', \beta < \beta' < \theta, \gamma < \gamma' < \theta$. Then we can rewrite, in the sense of distribution, $f = f_1 + f_2$, where

$$f_1(x) = \sum_{k=1}^{\infty} \widetilde{D}_k D_k(f)(x)$$

and

$$f_2(x) = \sum_{k=-\infty}^{0} \int [\widetilde{D}_k(x,y) - \widetilde{D}_k(x_0,y)]D_k(f)(y)d\mu(y)$$

where x_0 is any fixed point in X.

The condition $\|D_k(f)\|_\infty \le C2^{-ks}$ implies immediately that f_1 is continuous and bounded on X. The smoothness condition of wavelets $\widetilde{D}_k(x,y)$ yields

$$|\int [\widetilde{D}_k(x,y) - \widetilde{D}_k(x_0,y)]D_k(f)(y)d\mu(y)| \le C2^{(\epsilon-s)k}\rho^\epsilon(x,x_0)$$

where ϵ is the order of the regularity of wavelets $\widetilde{D}_k(x,y)$ and $0 < s < \epsilon < \theta$.

This implies that $f_2(x)$ is continuous on any compact subset in X, and thus $f(x)$ is continuous on any compact subset in X. To see f belongs to $\dot{C}^s$, we have

$$f(x) - f(x') = \sum_k \int [\widetilde{D}_k(x,y) - \widetilde{D}_k(x',y)]D_k(f)(y)d\mu(y).$$

We define a positive integer m by $2^{-m} \le \rho(x,x') < 2^{-m+1}$ and divide the above series into $\sum_{k\le m}$ and $\sum_{k>m}$.

For the first sum, the smoothness of wavelets $\widetilde{D}_k(x,y)$ and the size condition on $D_k(f)$ give

$$|\int [\widetilde{D}_k(x,y) - \widetilde{D}_k(x',y)]D_k(f)(y)d\mu(y)| \le C\rho^\epsilon(x,x')2^{(\epsilon-s)k}.$$

The contribution of the corresponding terms is $C\rho^\epsilon(x,x')2^{(\epsilon-s)m} \le C \rho^s(x,x')$. The size condition of wavelets $\widetilde{D}_k(x,y)$ and the size condition on $D_k(f)$ imply

$$|\int [\widetilde{D}_k(x,y) - \widetilde{D}_k(x',y)]D_k(f)(y)d\mu(y)| \le C2^{-sk}$$

which yields that the second sum is dominated by $C2^{-sm} \le C\rho^s(x,x')$.

The next theorem characterizes the homogeneous Hölder space $\dot{C}^s, 0 < s < \theta$, in terms of the order of magnitude of wavelet coefficients. To this end, we choose a family of wavelets ψ_λ with the order of the regularity $\epsilon, 0 < s < \epsilon < \theta$, as given in Theorem 3.25.

Theorem 4.4. *A wavelet series* $\sum\limits_\lambda \alpha(\lambda)\widetilde{\psi}_\lambda$ *defines a function in the homogeneous Hölder space* $\dot{C}^s, 0 < s < \theta$, *if and only if there exists a constant* C *such that, for all* $k \in \mathbb{Z}$ *and all* $\lambda = (k, \tau, \nu), \tau \in I_k, 1 \le \nu \le N(k, \tau)$,

$$|\alpha(\lambda)| \le C 2^{-\frac{1}{2}k} 2^{-sk}. \tag{4.9}$$

To see this theorem, we denote by $f_k(x)$ the series $\sum\limits_{\lambda \in \Lambda_k} \alpha(\lambda)\widetilde{\psi}_\lambda(x)$. This series is convergent because the wavelets $\widetilde{\psi}_\lambda$ are localized. But, however, the series $\sum\limits_k f_k(x)$ cannot converge in the usual sense, and must be renormalized as in Lemma 4.3. More precisely, by the regularity of wavelets $\widetilde{\psi}_\lambda(x)$ and the size condition on $\alpha(\lambda)$, for any fixed point $x_0 \in X$, we have

$$\begin{aligned}|f_k(x) - f_k(x_0)| &\le \sum_{\lambda \in \Lambda_k} |\alpha(\lambda)||[\widetilde{\psi}_\lambda(x) - \widetilde{\psi}_\lambda(x_0)]| \\ &\le C\rho^\epsilon(x, x_0) 2^{k(\epsilon - s)}\end{aligned}$$

where $\epsilon, s < \epsilon < \theta$, is the order of the regularity of wavelets $\widetilde{\psi}_\lambda$.

We divide the series $\sum\limits_k f_k(x)$ into the series $\sum\limits_{k>0} f_k$ and $\sum\limits_{k \le 0} f_k$. Then the first series converges and while the second is convergent on any compact subset in X. Finally, to show that the series $\sum\limits_k f_k(x)$ belongs to the homogeneous Hölder space $\dot{C}^s, 0 < s < \theta$, for any fixed $x, x' \in X$, we define an integer m by $2^{-m} \le \rho(x, x') < 2^{-m+1}$ and divide the series $\sum\limits_k f_k(x)$ into $\sum\limits_{k \le m}$ and $\sum\limits_{k > m}$. By the same method we used in Lemma 4.3, these imply the desired result.

We now return to the second kind of Hölder space, namely the inhomogeneous Hölder space $C^s(X)$. The difference between these two spaces is that the boundedness is needed for the inhomogeneous Hölder space $C^s(X), 0 < s < \theta$. To be more precise, the norm of $f \in C^s(X)$ is defined by $\|f\|_\infty + \sup\limits_{0 < h \le 1} \omega(h)h^s$, where $\omega(h)$ are modulus of continuity defined for the homogeneous Hölder space $\dot{C}^s(X)$. It is easy to see that the inhomogeneous Hölder space $C^s(X)$ is a Banach space.

It is impossible to characterize the inhomogeneous Hölder space $C^s(X)$ by the wavelet expansion given in Theorem 3.25, as we did for homogeneous Hölder space $\dot{C}^s(X)$ because $L^\infty(X)$ is a component of $C^s(X)$ and $L^\infty(X)$ cannot be characterized by the wavelet coefficients given in Theorem 3.25. Instead of using a wavelet expansion in Theorem 3.25, instead one uses a wavelet expansion given in Theorem 3.27. This is the following theorem.

Theorem 4.5. *Suppose that families of $\phi_\lambda, \widetilde{\phi}_\lambda$ are scaling functions and families of $\psi_\lambda, \widetilde{\psi}_\lambda, \widetilde{\widetilde{\psi}}_\lambda$ are wavelets given in Theorem 3.27 with the regularity epsilon, $0 < s < \epsilon < \theta$. A function $f \in L^1_{loc}(X)$ belongs to $C^s(X), 0 < s < \epsilon < \theta$ if and only if the wavelet coefficients*

$$\beta(\lambda) = \langle f, \widetilde{\phi}_\lambda \rangle$$

for $\lambda = (0, \tau, \nu)$, and

$$\alpha(\lambda) = \langle f, \widetilde{\psi}_\lambda \rangle$$

for $\lambda = (k, \tau, \nu), 1 \leq k \leq N$,

$$\alpha(\lambda) = \langle f, \psi_\lambda \rangle$$

for $\lambda = (k, \tau, \nu), k > N$, satisfy

$$|\beta(\lambda)| \leq C_0,$$

and

$$|\alpha(\lambda)| \leq C_1 2^{-\frac{k}{2}} 2^{-sk}.$$

The condition $|\beta(\lambda)| \leq C_0$ follows immediately from the fact that $f \in L^\infty(X)$ and $\widetilde{\phi} \in L^1(X)$. The conditions on the $\alpha(\lambda)$ follow from the same proof for homogeneous case because $\|f\|_\infty$ is not needed. We leave these details to the reader.

4.4 Lebesgue and Generalized Sobolev Spaces

We first characterize $L^p, 1 < p < \infty$, in terms of the wavelet coefficients.

Suppose that ψ_λ and $\widetilde{\psi}_\lambda$ are wavelets as given in Theorem 3.25. If $f \in L^p, 1 < p < \infty$, and $\alpha(\lambda)$ are the wavelet coefficients of f corresponding to ψ_λ, we have

$$f = \sum_\lambda \alpha(\lambda) \widetilde{\psi}_\lambda$$

where $\alpha(\lambda)$ are the wavelet coefficients of f and the series converges in the norm of L^p.

The characterization of $L^p, 1 < p < \infty$, is the following theorem.

Theorem 4.6. *For $1 < p < \infty$, the norms $\|f\|_p$, $\Big\|\Big\{\sum_\lambda |\alpha(\lambda)|^2 |\widetilde{\psi}_\lambda(x)|^2\Big\}^{\frac{1}{2}}\Big\|_p$, and $\Big\|\Big\{\sum_{\lambda\in\Lambda} |\alpha(\lambda)|^2 (\mu(Q_\lambda))^{-1} \chi_{Q_\lambda}\Big\}^{\frac{1}{2}}\Big\|_p$ are equivalent.*

Following the method given by Y. Meyer in [M2], we first prove that the first and the second norm are equivalent. We let Ω denote the product set

$\{-1,1\}^{\Lambda}$ and $d\mu(\omega)$ denote the Bernoulli probability measure, obtained by taking the product of the measures on each factor which gives a mass of $\frac{1}{2}$ to each of the points -1 and 1. An element ω of Ω is a sequence $\omega(\lambda), \lambda \in \Lambda$, consisting of -1 or 1.

For each $\omega \in \Omega$ we define the operator $T_\omega : L^2(X) \to L^2(X)$ given by $T_\omega(\psi_\lambda) = \omega(\lambda)\psi_\lambda$. We then get

Lemma 4.7. *The set $\{T_\omega : \omega \in \Omega\}$ is a bounded set of Calderón-Zygmund operators.*

This lemma is obvious, because the operators T_ω are bounded on $L^2(X)$ uniformly in ω and the kernels $K_\omega(x,y) = \sum_{\lambda\in\Lambda} \omega(\lambda)\widetilde{\psi}_\lambda(x)\overline{\psi}_\lambda(y)$ satisfy

$$|K_\omega(x,y)| \leq C\rho^{-1}(x,y)$$

and

$$|K_\omega(x,y) - K_\omega(x',y)| + |K_\omega(y,x) - K_\omega(y,x')| \leq C\rho^{-1-\epsilon}(x,y)$$

uniformly in ω for $\rho(x,x') \leq \frac{1}{2A}\rho(x,y)$ and some $\epsilon > 0$.

We also need Khinchin's well-known inequality ([Z]).

Lemma 4.8. *For $1 < p < \infty$, all the $L^p(\Omega, d\mu(\omega))$ norms are equivalent on the closed subspace of $L^2(\Omega)$ consisting of the functions $S(\omega) = \sum_{\lambda\in\Lambda} \langle f, \psi_\lambda\rangle\omega(\lambda)$. Moreover, for each p, there are constants $C_p \geq C'_p > 0$ such that*

$$C'_p\Big(\sum_\lambda |\langle f,\psi_\lambda\rangle|^2\Big)^{\frac{1}{2}} \leq \Big(\int_\Omega |S(\omega)|^p d\mu(\omega)\Big)^{\frac{1}{p}} \leq C_p\Big(\sum_\lambda |\langle f,\psi_\lambda\rangle|^2\Big)^{\frac{1}{2}}.$$

We now use Lemma 4.8 to show Theorem 4.6. By Calderón-Zygmund theory, $\|T_\omega(f)\|_p \leq C\|f\|_p$. Raising this inequality to the power p and then taking the mean over $\omega \in \Omega$ of the resulting inequality, we obtain a double integral over $X \times \Omega$, with respect to $d\mu(x)d\mu(\omega)$. Applying Fubini's theorem and, for each $x \in X$, Khinchin's inequality yields

$$\Big\|\Big\{\sum_\lambda |\alpha(\lambda)|^2|\widetilde{\psi}_\lambda(x)|^2\Big\}^{\frac{1}{2}}\Big\|_p \leq C'\|f\|_p.$$

To get the converse inequality, we have

$$C_p^{-p}\int_{\omega\in\Omega} \|T_\omega(f)\|_p^p d\mu(\omega) \leq \Big\|\Big\{\sum_\lambda |\alpha(\lambda)|^2|\widetilde{\psi}_\lambda(x)|^2\Big\}^{\frac{1}{2}}\Big\|_p^p$$

where the second part of Khinchin's inequality is used.

Finally, observing that $T_\omega^2 = I$, we have $\|f\|_p \leq C\|T_\omega(f)\|_p$, for each ω, and this gives the converse part of the equivalence between the first and the second norm in Theorem 4.6.

We now prove that the first and the third norm are equivalent. Suppose that $\{D_k\}_{k\in\mathbb{Z}}$ are given as in Section 1.3. We claim that

$$\left\|\left\{\sum_{\lambda\in\Lambda}|\langle f,\psi_\lambda\rangle|^2(\mu(Q_\lambda))^{-1}\chi_{Q_\lambda}\right\}^{\frac{1}{2}}\right\|_p \leq C\left\|\left\{\sum_k |D_k(f)|^2\right\}^{\frac{1}{2}}\right\|_p \leq C\|f\|_p$$

where the last inequality follows from the Littlewood-Paley estimates on $L^p, 1 < p < \infty$, given by G. David, J. L. Journé and S. Semmes. This implies that the third norm is dominated by the first norm.

The proof that the first norm is dominated by the third norm follows from a standard method. Indeed, the claim and the duality argument yield

$$\begin{aligned}\|f\|_p &= \sup_{\|h\|_{p'}\leq 1}\left|\left\langle \sum_\lambda \langle f,\psi_\lambda\rangle\widetilde{\psi}_\lambda, h\right\rangle\right| \\ &\leq C \sup_{\|h\|_{p'}\leq 1}\int\sum_\lambda |\langle f,\psi_\lambda\rangle||\langle\widetilde{\psi}_\lambda,h\rangle|(\mu(Q_\lambda))^{-1}\chi_{Q_\lambda}(x)d\mu(x) \\ &\leq C\left\|\left\{\sum_{\lambda\in\Lambda}|\langle f,\psi_\lambda\rangle|^2(\mu(Q_\lambda))^{-1}\chi_{Q_\lambda}\right\}^{\frac{1}{2}}\right\|_p\end{aligned}$$

where the last inequality follows from the claim.

Finally, we prove the claim. By the construction of wavelets given in Theorem 3.25, $\psi_\lambda(x) = \sqrt{\mu(Q_\lambda)}D_k(y_\lambda, x), \lambda \in \Lambda = \{(k,\tau,\nu) : k \in \mathbb{Z}, \tau \in I_k, 1 \leq \nu \leq N(k,\tau)\}$, where $D_k(x,y)$ are mother functions and the sampling grid y_λ is given by an arbitrary point in Q_λ. Thus, be choosing y_λ, we have

$$\langle f,\psi_\lambda\rangle\chi_{Q_\tau^{k,\nu}}(x) \leq \sqrt{\mu(Q_\lambda)}D_k(f)(x)\chi_{Q_\tau^{k,\nu}}(x).$$

Taking power by 2, then Multiplying $(\mu(Q_\lambda))^{-1}$ for the both sides and finally summing up over λ give

$$\sum_{\lambda\in\Lambda}|\langle f,\psi_\lambda\rangle|^2(\mu(Q_\lambda))^{-1}\chi_{Q_\lambda}(x) \leq C\sum_k |D_k(f)(x)|^2$$

which implies immediately the claim.

We remark that Theorem 4.6, based on Theorem 4.1, holds for any choice of wavelet expansions.

We now intend to study the generalized Sobolev spaces on a space of homogeneous type. Let us recall these spaces on standard case of $\mathbb{R}^n$. There are two kinds of Sobolev spaces on $\mathbb{R}^n$, that is, the homogeneous Sobolev space $\dot{L}^{p,s}(\mathbb{R}^n)$ and the inhomogeneous Sobolev space $L^{p,s}(\mathbb{R}^n)$.

If $1 < p < \infty$ and $s \geq 0$ is a positive real number, then $L^{p,s}(\mathbb{R}^n)$ is the subspace of L^p consisting of functions $f \in L^p$ such that $(I - \Delta)^{\frac{s}{2}} f$ is also in L^p, where Δ is the Laplacian on $\mathbb{R}^n$.

The homogeneous Sobolev space corresponding to $\dot{L}^{p,s}(\mathbb{R}^n)$ is formally defined by $\|(-\Delta)^{\frac{s}{2}} f\|_p < \infty$. See [M2] for more discussion.

Using the Littlewood-Paley analysis, the generalized Sobolev space on spaces of homogeneous type had been defined in [HS] and [H2]. We only describe and study the homogeneous case. The inhomogeneous case is similar and we leave these details to the reader.

Definition 4.9. Suppose that $\{D_k\}_{k\in\mathbb{Z}}$ are mother functions as given in Theorem 3.15. We say that $f \in (\dot{\mathcal{M}}_0(\beta,\gamma))'$ belongs to the generalized homogeneous Sobolev space $\dot{L}^{p,s}(X), -\theta < s < \theta, \max\{\frac{1}{1+\theta}, \frac{1}{1+\theta+s}\} < p < \infty$, if

$$\|f\|_{\dot{L}^{p,s}} = \left\| \left\{ \sum_k (2^{sk}|D_k(f)|)^2 \right\}^{\frac{1}{2}} \right\|_p < \infty.$$

Suppose again that ψ_λ and $\widetilde{\psi}_\lambda$ are wavelets with the regularity order $\epsilon, \epsilon < \theta$, and $f \in (\dot{\mathcal{M}}_0(\beta,\gamma))', 0 < \beta,\gamma < \epsilon < \theta$. By Theorem 3.25, we have

$$f = \sum_\lambda \langle f, \psi_\lambda \rangle \widetilde{\psi}_\lambda$$

where the series converges in the sense of distribution.

Then we have the following criterion for $\dot{L}^{p,s}(X)$ in terms of the wavelet coefficients.

Theorem 4.10. *Let* $-\theta < s < \theta, \max\{\frac{1}{1+\theta}, \frac{1}{1+\theta+s}\} < p < \infty$. *Suppose that a distribution* $f \in (\dot{\mathcal{M}}_0(\beta,\gamma))', 0 < \beta,\gamma < \epsilon < \theta$ *and* f *has a wavelet expansion* $f = \sum_\lambda \langle f, \psi_\lambda \rangle \widetilde{\psi}_\lambda$, *where the regularity order of* ψ_λ *is* $\epsilon, |s| < \epsilon < \theta, \max(\frac{1}{1+\epsilon}, \frac{1}{1+\epsilon+s}) < p, 0 < \beta,\gamma < \epsilon$. *Then* f *belongs to* $\dot{L}^{p,s}(X)$ *if and only if*

$$\left\{ \sum_{\lambda\in\Lambda} \left[2^{k(s+\frac{1}{2})} |\langle f, \psi_\lambda \rangle| \chi_{Q_\lambda} \right]^2 \right\}^{\frac{1}{2}} \in L^p(X).$$

The proof is similar to one given in Theorem 4.6 for the equivalence between the first and the third norm. Indeed, we first choose the point $y_\lambda \in Q_\lambda$ so that

$$|\langle f, \psi_\lambda \rangle| \chi_{Q_\lambda}(x) = \sqrt{\mu(Q_\lambda)} |D_k(f)(y_\lambda)| \chi_{Q_\lambda}(x) \leq C 2^{\frac{-k}{2}} |D_k(f)(x)| \chi_{Q_\lambda}(x).$$

Taking power by 2, multiplying $2^{k(s+\frac{1}{2})}$ and then summing over λ yield one implication:

$$\left\|\left\{\sum_{\lambda\in\Lambda}\left[2^{k(s+\frac{1}{2})}|\langle f,\psi_\lambda\rangle|\chi_{Q_\lambda}\right]^2\right\}^{\frac{1}{2}}\right\|_p\le C\left\|\left\{\sum_k(2^{sk}|D_k(f)|)^2\right\}^{\frac{1}{2}}\right\|_p\le C\|f\|_{\dot{L}^{p,s}}.$$

Instead one chooses the point $y_\lambda\in Q_\lambda$ so that

$$|\langle f,\psi_\lambda\rangle|\chi_{Q_\lambda}(x)=\sqrt{\mu(Q_\lambda)}|D_k(f)(y_\lambda)|\chi_{Q_\lambda}(x)\ge C2^{\frac{-k}{2}}|D_k(f)(x)|\chi_{Q_\lambda}(x)$$

which implies

$$\left\|\left\{\sum_k(2^{sk}|D_k(f)|)^2\right\}^{\frac{1}{2}}\right\|_p\le C\left\|\left\{\sum_{\lambda\in\Lambda}\left[2^{k(s+\frac{1}{2})}|\langle f,\psi_\lambda\rangle|\chi_{Q_\lambda}\right]^2\right\}^{\frac{1}{2}}\right\|_p$$

and from this the other implication follows.

4.5 Wavelets, the Hardy and BMO Spaces

In the previous section, the extreme cases $p=1$ and $p=\infty$ are excluded. This is because the spaces L^1 or L^∞ on $\mathbb{R}^n$ have no unconditional bases and hence it is impossible to characterize these spaces by size properties on wavelet coefficients. But there are good substitutes for these two spaces, that is, the Hardy space $H^1(\mathbb{R}^n)$ and space of BMO. One should observe that the wavelets developed in Chapter 3 are not orthonormal bases instead they are wavelet frames. We, however, still can use such wavelet frames to study the Hardy space H^1 and BMO.

Following Y. Meyer in [M2], we first give some general definitions for series of vectors in a Banach space.

Let B be a Banach space and $\sum_0^\infty x_k$ a series of elements of B. We say that this series converges unconditionally to an element $x\in B$ if, for each $\epsilon>0$, there exists a finite set $F(\epsilon)\subset\mathbb{N}$ such that, for every finite set $F\subset\mathbb{N}$ containing $F(\epsilon)$,

$$\left\|\sum_{k\in F}x_k-x\right\|\le\epsilon.$$

We say that a series $\sum_0^\infty x_k$ is unconditionally convergent if there exists an x such that the series converges unconditionally to x. Therefore, if a series $\sum_0^\infty x_k$ is unconditionally convergent, there exists a constant C such that, for every integer $n\ge1$ and every sequence $\alpha_0,\alpha_1,\cdots,\alpha_n$ of real or complex numbers satisfying $|\alpha_0|\le1,\cdots,|\alpha_n|\le1$, we have

$$\|\alpha_0x_0+\cdots+\alpha_nx_n\|\le C.$$

The unconditional convergence of a series $\sum_0^\infty x_k$ can be stated by another way: the series is commutatively convergent, that is, for every permutation of the integers $\sigma : \mathbb{N} \to \mathbb{N}$, the series $\sum_0^\infty x_{\sigma k}$ converges in norm to $x = \sum_0^\infty x_k$.

In this section, we shall first consider the limiting case where $p = 1$. We will only consider certain functions $f \in L^1(X)$ having the special property that their wavelet series converges unconditionally to f. We will give a characterization of such functions in terms of the wavelet coefficients. This characterization differs from the traditional approach given by E. Stein and G. Weiss, and C. Fefferman and E. Stein in [SW] and [FS]. We will prove that all these characterizations are equivalent and give the first characterization of the space $H^1(X)$ now.

We say that $f \in H^1(X)$ if $f \in L^1(X)$ and the wavelet series $\sum_{\lambda \in \lambda} \langle f, \widetilde{\psi}_\lambda \rangle \psi_\lambda(x)$ converges unconditionally to f.

As we mentioned above, this imposes the existence of a constant $C \geq 1$ such that, for each finite subset $F \subset \Lambda$ and every sequence $\epsilon(\lambda), \lambda \in F$, taking the values -1 or 1, we have

$$\left\| \sum_{\lambda \in F} \epsilon(\lambda) \langle f, \widetilde{\psi}_\lambda \rangle \psi_\lambda(x) \right\|_1 \leq C. \tag{4.10}$$

Following [M2], the upper bound, over all F and sequences $\epsilon(\lambda), \lambda \in F$, of the left-hand side of (4.10) will be the first norm we consider on the space $H^1(X)$.

We need to specify which particular wavelets ψ_λ are being used. We recall the construction of Coifman. $\{S_k\}_{k \in \mathbb{Z}}$ is constructed as described in Section 1.3 of Chapter 1. Set $D_k = S_k - S_{k-1}$ and $\psi_\lambda(x) = \sqrt{\mu(Q_\lambda)} D_k(y_\lambda, x)$, where $\lambda \in \Lambda = \{(k, \tau, \nu) : k \in \mathbb{Z}, \tau \in I_k, 1 \leq \nu \leq N(k, \tau)\}$ and y_λ is chosen to be the center of Q_λ.

As in [M2], using Khinchin's well-known inequality: there exists a constant $C > 1$ such that, for every integer $n \geq 1$ and every sequence $a_1, \cdots, a_n$ of complex numbers,

$$\left(\sum_1^n |a_j|^2 \right)^{\frac{1}{2}} \leq C 2^{-n} \sum_{\epsilon_1} \cdots \sum_{\epsilon_n} |\epsilon_1 a_1 + \cdots + \epsilon_n a_n| \tag{4.11}$$

where, on the right-hand side, the sum is taken over all sequences $\epsilon = (\epsilon_1, \cdots, \epsilon_n)$ of 1s and -1s.

The inequality implies the existence of a constant C such that

$$\int_X \left(\sum_{\lambda \in F} |\langle f, \widetilde{\psi}_\lambda \rangle|^2 |\psi_\lambda(x)|^2 \right)^{\frac{1}{2}} d\mu(x) \leq C$$

for every finite subset $F \subset \Lambda$.

This leads to the second norm of $H^1(X)$, defined by

$$\int\limits_X \left(\sum_{\lambda \in \Lambda} |\langle f, \widetilde{\psi}_\lambda \rangle|^2 |\psi_\lambda(x)|^2 \right)^{\frac{1}{2}} d\mu(x) < \infty. \tag{4.12}$$

As explained in [M2], it is not obvious that this second norm is equivalent to the first, because, in passing from (4.10) to (4.12), we have lost information since we have replaced an inequality which is uniform, with respect to sequences of 1 and -1, by an inequality in the mean. But, however, we will prove that these two norms are equivalent.

To show this, we will use other three norms of $H^1(X)$. First, by the construction of ψ_λ, it is easy to see that there are two constants $C_1 > 0$ and $\gamma > 0$ such that, for every $\lambda \in \Lambda$, $|\psi_\lambda(x)| \geq C_1$ when $x \in Q_{\lambda,\gamma}$, where $Q_{\lambda,\gamma} \subset Q_\lambda$ with $\mu(Q_{\lambda,\gamma}) \geq \gamma\mu(Q_\lambda)$.

Then the third norm of $H^1(X)$ will be

$$\left\| \left(\sum_{\lambda \in \Lambda} |\langle f, \widetilde{\psi}_\lambda \rangle|^2 \mu(Q_{\lambda,\gamma})^{-1} \chi_{Q_{\lambda,\gamma}} \right)^{\frac{1}{2}} \right\|_1 \tag{4.13}$$

where $\chi_{Q_{\lambda,\gamma}}$ is the characteristic function of $Q_{\lambda,\gamma}$.

The only relevant property of $Q_{\lambda,\gamma}$ is the fact that $\mu(Q_{\lambda,\gamma}) \geq \gamma\mu(Q_\lambda)$.

The fourth norm we use is defined by (4.13), but with $Q_{\lambda,\gamma}$ replaced by Q_λ.

The last norm of $H^1(X)$ does not involve wavelet expansions. It is the well-known atomic definition of $H^1(X)$.

Definition 4.11. An atom of $H^1(X)$ is a function $a(x)$, belonging to $L^2(X)$, such that there exists a ball $B \in X$, whose volume is denoted by $\mu(B)$, and for which the three following properties hold:

$$a(x) = 0 \tag{4.14}$$

if $x \notin B$;

$$\|a\|_2 \leq \mu(B)^{\frac{-1}{2}}; \tag{4.15}$$

$$\int\limits_B a(x) d\mu(x) = 0. \tag{4.16}$$

Applying the Cauchy-Schwarz inequality, we see that $a(x)$ is integrable and that $\|a\|_1 \leq 1$, so that property (4.16) makes sense.

We say that a function $f \in L^1(X)$ belongs to atomic $H^1(X)$ if there exists a sequence $a_j(x)$ of atoms and a sequence λ_j of scaler coefficients such that

$f(x) = \sum_0^\infty \lambda_j a_j(x)$ with $\sum_0^\infty |\lambda_j| < \infty$, where the series clearly converges in $L^1(X)$.

The norm of atomic $f \in H^1(X)$ is then defined as the infimum of the quantities $\sum_0^\infty |\lambda_j|$ corresponding to all possible atomic decompositions of f.

The main result of this section is the following theorem.

Theorem 4.12. *The above five norms of $H^1(X)$ are equivalent.*

To clarify the proof, we denote the following properties whose equivalence will be proved, by A, B, C, D and E.

(A) $\sup_{F\subset\lambda} \sup_{\epsilon(\lambda)=\pm 1} \left\| \sum_{\lambda\in F} \epsilon(\lambda)\alpha(\lambda)\psi_\lambda(x)\right\|_1 < \infty$;

(B) $\left\|\left(\sum_{\lambda\in\Lambda} |\alpha(\lambda)|^2|\psi_\lambda|^2\right)^{\frac{1}{2}}\right\|_1 < \infty$;

(C) $\left\|\left(\sum_{\lambda\in\Lambda} |\alpha(\lambda)|^2\mu(Q_{\lambda,\gamma})^{-1}\chi_{Q_{\lambda,\gamma}}\right)^{\frac{1}{2}}\right\|_1 < \infty$;

(D) $\left\|\left(\sum_{\lambda\in\Lambda} |\alpha(\lambda)|^2\mu(Q_{\lambda})^{-1}\chi_{Q_{\lambda}}\right)^{\frac{1}{2}}\right\|_1 < \infty$; and

(E) $f(x) = \sum_\lambda \alpha(\lambda)\psi_\lambda(x)$ has an atomic decomposition. Here $\alpha(\lambda) = \langle f, \widetilde{\psi}_\lambda\rangle$ are the wavelet coefficients of f.

We have already explained that $(A) \Rightarrow (B)$, and it is clear that $(B) \Rightarrow (C)$. Note that if $\mu(Q_{\lambda,\gamma}) > \gamma\mu(Q_\lambda)$ then for all $r > 0, \chi_{Q_\lambda} \le \gamma^{-\frac{1}{r}}(M(\chi_{Q_{\lambda,\gamma}})^r)^{\frac{1}{r}}$, where M is the Hardy-Littlewood maximal operator. Choosing $r < 1$ and applying the Fefferman -Stein vector-valued maximal function inequality yields

$$\begin{aligned}
&\left\|\left(\sum_{\lambda\in\Lambda} |\alpha(\lambda)|^2\mu(Q_{\lambda})^{-1}\chi_{Q_{\lambda}}\right)^{\frac{1}{2}}\right\|_1 \\
&\qquad \le C\gamma^{-\frac{1}{r}}\left\|\left(\sum_{\lambda\in\Lambda} (M(|\alpha(\lambda)|\mu(Q_\lambda,\gamma)^{-\frac{1}{2}}\chi_{Q_{\lambda,\gamma}})^r)^{\frac{2}{r}}\right)^{\frac{r}{2}}\right\|_{\frac{1}{r}}^{\frac{1}{r}} \\
&\qquad \le C\left\|\left(\sum_{\lambda\in\Lambda} |\alpha(\lambda)|^2\mu(Q_{\lambda,\gamma})^{-1}\chi_{Q_{\lambda,\gamma}}\right)^{\frac{1}{2}}\right\|_1
\end{aligned}$$

which yields $(C) \Rightarrow (D)$.

We now prove $(D) \Rightarrow (E)$. Let us denote

$$S(\alpha)(x) = \left(\sum_{\lambda} |\alpha(\lambda)|^2\mu(Q_{\lambda})^{-1}\chi_{Q_{\lambda}}(x)\right)^{\frac{1}{2}}.$$

Set $\Omega_k = \{x \in X : S(\alpha)(x) > 2^k\}$ and $B_k = \{Q : \mu(Q\cap\Omega_k) > \frac{1}{2}\mu(Q), \mu(Q\cap\Omega_{k+1}) \le \frac{1}{2}\mu(Q)\}$. Therefore, we can write

$$\sum_\lambda \alpha(\lambda)\psi_\lambda(x) = \sum_k \sum_{Q_\lambda \in B_k} \alpha(\lambda)\psi_\lambda(x) = \sum_k \sum_{\widetilde{Q}\in B_k} \sum_{Q_\lambda\subset \widetilde{Q}, Q_\lambda\in B_k} \alpha(\lambda)\psi_\lambda(x)$$

where $\widetilde{Q}$ are maximal dyadic cubes in B_k, that is, if both $\widetilde{Q}$ and Q are in B_k and $\widetilde{Q}\subset Q$, then $\widetilde{Q}=Q$.

Let us denote that $a_k(\widetilde{Q})(x) = (C\mu(\widetilde{Q}) \sum_{Q_\lambda\subset \widetilde{Q}, Q_\lambda\in B_k} |\alpha(\lambda)|^2)^{\frac{-1}{2}} \sum_{Q_\lambda\subset \widetilde{Q}, Q_\lambda\in B_k} \alpha(\lambda)\psi_\lambda(x)$ and $\lambda_k(\widetilde{Q}) = (C\mu(\widetilde{Q}) \sum_{Q_\lambda\subset \widetilde{Q}, Q_\lambda\in B_k} |\alpha(\lambda)|^2)^{\frac{1}{2}}$, where C is a constant to be chosen later. Then

$$\sum_\lambda \alpha(\lambda)\psi_\lambda(x) = \sum_k \sum_{\widetilde{Q}\in B_k} \lambda_k(\widetilde{Q}) a_k(\widetilde{Q})(x).$$

To see that this gives an atomic decomposition, we clearly have that (i) each $a_k(\widetilde{Q})(x)$ has the support $\widetilde{Q}$; (ii) by duality argument,

$$\begin{aligned}
\Big\| \sum_{Q_\lambda\subset \widetilde{Q}, Q_\lambda\in B_k} \alpha(\lambda)\psi_\lambda(x)\Big\|_2 &= \sup_{\|h\|_2\le 2} \Big|\Big\langle \sum_{Q_\lambda\subset \widetilde{Q}, Q_\lambda\in B_k} \alpha(\lambda)\psi_\lambda(x), h\Big\rangle\Big| \\
&\le C \sup_{\|h\|_2\le 2} \Big(\sum_{Q_\lambda\subset \widetilde{Q}, Q_\lambda\in B_k} |\alpha(\lambda)|^2\Big)^{\frac{1}{2}} \|h\|_2 \\
&\le C\Big(\sum_{Q_\lambda\subset \widetilde{Q}, Q_\lambda\in B_k} |\alpha(\lambda)|^2\Big)^{\frac{1}{2}}
\end{aligned}$$

which implies $\|a_k(\widetilde{Q})(x)\|_2 \le \mu(\widetilde{Q})^{-\frac{1}{2}}$; (iii) finally,

$$\begin{aligned}
\sum_k \sum_{\widetilde{Q}\in B_k} |\lambda_k(\widetilde{Q})| &\le C\sum_k \sum_{\widetilde{Q}\in B_k} \mu(\widetilde{Q})^{\frac{1}{2}} \sum_{Q_\lambda\subset \widetilde{Q}, Q_\lambda\in B_k} (|\alpha(\lambda)|^2)^{\frac{1}{2}} \qquad (4.17)\\
&\le C\sum_k \Big\{ \sum_{\widetilde{Q}\in B_k} \mu(\widetilde{Q})\Big\}^{\frac{1}{2}} \Big\{ \sum_{\widetilde{Q}\in B_k} \sum_{Q_\lambda\subset \widetilde{Q}, Q_\lambda\in B_k} (|\alpha(\lambda)|^2)\Big\}^{\frac{1}{2}}.
\end{aligned}$$

To estimate the last term above, let us denote $\widetilde{\Omega}_k = \{x\in X : M(\chi_{\Omega_k})(x) > \frac{1}{2}\}$. It is clear that if $Q_\lambda \in B_k$, then $Q_\lambda \subset \widetilde{\Omega}_k$. Therefore the last term in (4.17) is dominated by

$$\begin{aligned}
&C\sum_k \mu(\widetilde{\Omega}_k)^{\frac{1}{2}} \Big\{ \sum_{Q_\lambda\in B_k} |\alpha(\lambda)|^2)\Big\}^{\frac{1}{2}} \\
&\le C\sum_k \mu(\widetilde{\Omega}_k)^{\frac{1}{2}} 2^k \mu(\widetilde{\Omega}_k)^{\frac{1}{2}} \\
&\le C\sum_k 2^k \mu(\widetilde{\Omega}_k) \le C\sum_k 2^k \mu(\Omega_k) \le C\|S(\alpha)\|_1 < \infty \qquad (4.18)
\end{aligned}$$

where the first inequality follows from the facts that if $Q \in B_k$, then $Q \subset \widetilde{\Omega}_k$ and $\sum_{Q_\lambda \in B_k} |\alpha(\lambda)|^2 \leq C \int_{\widetilde{\Omega}_k \setminus \Omega_k} S^2(\alpha(\lambda)(x) d\mu(x) \leq C 2^{2k} \mu(\widetilde{\Omega}_k)$, and the third inequality follows from the fact that $\mu(\widetilde{\Omega}_k) \leq C\mu(\Omega_k)$, by the maximal operator theorem.

Finally, the verification of $(E) \Rightarrow (D)$ is an easy exercise, because it is sufficient to do it for each atom taken separately. The same proof yields $(E) \Rightarrow (A)$. The proof of Theorem 4.12 is complete.

Before we continue to study the Hardy spaces $H^p(X)$, $p < 1$, we first describe the space BMO in terms of wavelet expansions. The space BMO has been defined in Section 1.2 of Chapter 1. For this purpose, we choose wavelet expansion $f(x) = \sum_\lambda \alpha(\lambda)\psi_\lambda(x), \alpha(\lambda) = \langle f, \widetilde{\psi}_\lambda \rangle$. Then we have

Theorem 4.13. *Let $b(x)$ be a function belonging to $BMO(X)$. Then its wavelet coefficients $\alpha(\lambda) = \langle b, \widetilde{\psi}_\lambda \rangle$ satisfy Carleson's condition, as follows: There exists a constant C such that for each dyadic cube Q,*

$$\sum_{Q_\lambda \subset Q} |\alpha(\lambda)|^2 \leq C\mu(Q). \tag{4.19}$$

Conversely, if the coefficients $\alpha(\lambda), \lambda \in \Lambda$, satisfy (4.19), then the wavelet expansion $\sum_\lambda \alpha(\lambda)\psi_\lambda(x)$ converges, in the $\sigma(BMO, H^1)$–topology, to a function of BMO.

We first prove that if $b \in BMO$, then (4.19) holds.

For given Q, we decompose $b(x)$ by $b(x) = b_1(x) + b_2(x) + c(Q)$, where $c(Q)$ is the mean of $b(x)$ over mQ and where $b_1(x) = b(x) - c(Q)$ if $x \in mQ$, and $b_1(x) = 0$ otherwise. We obtain

$$\sum_{Q_\lambda \subset Q} |\langle b_1, \widetilde{\psi}_\lambda \rangle|^2 \leq \sum_{\lambda \in \Lambda} |\langle b_1, \widetilde{\psi}_\lambda \rangle|^2 \leq C\|b_1\|_2^2 \leq Cm\|b\|_{BMO}^2 \mu(Q).$$

By the cancellation of $\widetilde{\psi}_\lambda(x)$, we may assume the average of b_2 over Q is zero because otherwise one can replace b_2 by $b_2 - (b_2)_Q$, where $(b_2)_Q = \frac{1}{\mu(Q)} \int_Q b_2(x) d\mu(x)$. Then we have

$$|\langle b_2, \widetilde{\psi}_\lambda \rangle| \leq C 2^{(k_0 - k)\epsilon} 2^{-\frac{1}{2}k} \|b\|_{BMO}$$

where $\lambda = (k, \tau, \nu)$ and $\mu(Q) \approx 2^{-k_0}$. This implies

$$\sum_{Q_\lambda \subset Q} |\langle b_2, \widetilde{\psi}_\lambda \rangle|^2 \leq C\|b\|_{BMO}^2 \sum_{k \geq k_0} 2^{2(k_0 - k)\epsilon} \mu(Q) \leq C\|b\|_{BMO}^2 \mu(Q).$$

Suppose that (4.19) is satisfied. Following [M2], let B be a ball with center x_0 and of radius $r > 0$. We define the integer $q \in \mathbb{Z}$ by $2^{-q} \leq r < 2^{-q+1}$. We split the sum $\sum_\lambda \alpha(\lambda)\psi_\lambda(x)$ into two parts. We first consider "small" cubes of side $2^{-k} \le 2^{-q}$ and then "large" cubes for which $k < q$. The wavelets corresponding to the small cubes are themselves of two kinds: their supports either meet B or don't meet B. If a small cube Q_λ has the property that mQ_λ meets B, then Q_λ is necessarily contained in MB, where $M > 1$ is a constant depending only on m. Let $b = b_1 + b_2$, where b_1 is corresponding to the small cubes and b_2 to the large cubes. Then b_1 splits into $b_{1,1} + b_{1,2}$, and $b_{1,2} = 0$ on B. Then, by the fact that the small cubes Q_λ are contained in MB,

$$\|b_{1,1}\|_2^2 \le \sum_{Q_\lambda \subset MB} |\alpha(\lambda)|^2 \le C\mu(B).$$

To deal with the large cubes and the corresponding subseries b_2 of $\sum_\lambda \alpha(\lambda)\psi_\lambda(x)$, by the regularity of wavelets, we have $|\psi_\lambda(x) - \psi_\lambda(x_0)| \le C2^{k(1+\epsilon)}2^{-\frac{k}{2}}\rho(x,x_0)^\epsilon$. Since $|\alpha(\lambda)| \le C(\mu(Q_\lambda))^{\frac{1}{2}} \le C2^{-\frac{k}{2}}$ by (4.19), summing up gives $\sum_{k<q} 2^{k\epsilon}\rho(x,x_0)^\epsilon = 2^{q\epsilon}\rho(x,x_0)^\epsilon \le 2^\epsilon$, since $\rho(x,x_0) \le r < 2^{-q+1}$, where x_0 is the center of B and $r > 0$ is its radius.

Using the above theorem, we prove

Theorem 4.14. *Let $b(x)$ be a function belonging to the space BMO. Then $b(x)$ defines a continuous linear functional on $H^1(X)$ by*

$$\ell(f) = \sum_\lambda \alpha(\lambda)\overline{\beta}(\lambda) \tag{4.20}$$

where $\alpha(\lambda) = \langle f, \widetilde{\psi}_\lambda\rangle$, $\beta(\lambda) = \langle b, \widetilde{\psi}_\lambda\rangle$.

Conversely, every continuous linear functional on $H^1(X)$ is defined in this way.

We use the first part of Theorem 4.13 to show the first part of Theorem 4.14. Let all notations be same as in Theorem 4.12. Then, by the same proof of the implication $(D) \Rightarrow (E)$ in Theorem 4.12, we have

$$\begin{aligned}
\left|\sum_\lambda \alpha(\lambda)\overline{\beta}(\lambda)\right| &= \left|\sum_k \sum_{Q_\lambda \in B_k} \alpha(\lambda)\overline{\beta}(\lambda)\right| \\
&= \left|\sum_k \sum_{\widetilde{Q}\in B_k} \sum_{Q_\lambda \subset \widetilde{Q}, Q_\lambda \in B_k} \alpha(\lambda)\overline{\beta}(\lambda)\right| \\
&\le \sum_k \sum_{\widetilde{Q}\in B_k} \left\{\sum_{Q_\lambda \subset \widetilde{Q}, Q_\lambda \in B_k} |\alpha(\lambda)|^2\right\}^{\frac{1}{2}} \left\{\sum_{Q_\lambda \subset \widetilde{Q}, Q_\lambda \in B_k} |\beta(\lambda)|^2\right\}^{\frac{1}{2}} \\
&\le C\|b\|_{BMO} \sum_k \sum_{\widetilde{Q}\in B_k} \mu(\widetilde{Q})^{\frac{1}{2}} \left\{\sum_{Q_\lambda \subset \widetilde{Q}, Q_\lambda \in B_k} |\alpha(\lambda)|^2\right\}^{\frac{1}{2}}
\end{aligned}$$

since, by the first part of Theorem 4.13,

$$\sum_{Q_\lambda \subset \widetilde{Q}} |\beta(\lambda)|^2 \leq C\mu(\widetilde{Q}).$$

This together with the same proof as in the implication $(D) \Rightarrow (E)$ in Theorem 4.12 gives

$$\left|\sum_\lambda \alpha(\lambda)\overline{\beta}(\lambda)\right| \leq C\|b\|_{BMO} \sum_k 2^k \mu(\Omega_k) \leq C\|b\|_{BMO}\|f\|_{H^1}$$

which implies the first part of Theorem 4.14.

To show the second part of Theorem 4.14, we define a sequence space by $\alpha(\lambda) \in s$ if $\left\{\sum\limits_\lambda |\alpha(\lambda)|^2 \mu(Q_\lambda)^{-1}\chi_{Q_\lambda}(x)\right\}^{\frac{1}{2}}$ belongs to $L^1(X)$ and this L^1 norm is defined as the norm of the sequence $\alpha(\lambda)$ in s. We also define another sequence space by $\beta(\lambda) \in c$ if, for each dyadic cube Q, $\sum\limits_{Q_\lambda \subset Q} |\beta(\lambda)|^2 \leq C\mu(Q)$. The norm of $\beta(\lambda) \in c$ is defined by the smallest constant C.

We then have

Theorem 4.15. $s^* = c$, *where* s^* *is the dual of* s.

Applying the similar proof of the first part of Theorem 4.14 implies that $c \subset s^*$. Conversely, every $\ell \in s^*$ is of the form $\ell(\alpha(\lambda)) = \sum\limits_\lambda \alpha(\lambda)\overline{\beta}(\lambda)$ for some sequence $\beta(\lambda)$. Let Q be fixed cube and define s_0 by the sequence space of all $\alpha(\lambda)$ so that $Q_\lambda \subset Q$ and let σ be a measure on s_0 such that the $\sigma-$measure at Q_λ is $\frac{\mu(Q_\lambda}{\mu(Q)}$. Then

$$\begin{aligned}
\left(\frac{1}{\mu(Q)}\sum_{Q_\lambda \subset Q}|\beta(\lambda)|^2\right)^{\frac{1}{2}} &= \|\mu(Q_\lambda)^{-\frac{1}{2}}\beta(\lambda)\|_{l^2(s_0,d\sigma)} \\
&= \sup_{\|\alpha(\lambda)\|_{l^2(s_0,d\sigma)}\leq 1}\left|\frac{1}{\mu(Q)}\sum_{Q_\lambda \subset Q}\alpha(\lambda)\mu(Q_\lambda)^{-\frac{1}{2}}\overline{\beta}(\lambda)\right| \\
&\leq \|\beta\|_{s^*}\sup_{\|\alpha(\lambda)\|_{l^2(s_0,d\sigma)}\leq 1}\left\|\alpha(\lambda)\frac{\mu(Q_\lambda)^{\frac{1}{2}}}{\mu(Q)}\right\|_s.
\end{aligned}$$

By Hölder's inequality,

$$\begin{aligned}
\left\|\alpha(\lambda)\frac{\mu(Q_\lambda)^{\frac{1}{2}}}{\mu(Q)}\right\|_s &= \frac{1}{\mu(Q)}\int_Q\left(\sum_{Q_\lambda \subset Q}|\alpha(\lambda)|^2\chi_{Q_\lambda}\right)^{\frac{1}{2}} \\
&\leq \left(\frac{1}{\mu(Q)}\int_Q\sum_{Q_\lambda \subset Q}|\alpha(\lambda)|^2\chi_{Q_\lambda}\right)^{\frac{1}{2}} = \|\alpha(\lambda)\|_{l^2(s_0,d\sigma)}.
\end{aligned}$$

We now return to the proof of the second part of Theorem 4.14. Suppose that ℓ is a continuous linear functional defined on $H^1(X)$. We define $l(\alpha(\lambda)) = \ell(f)$, where $f \in H^1(X), \alpha(\lambda)$ are the wavelet coefficients of f. Then, by Theorem 4.12 and Theorem 4.13, l is a continuous linear functional defined on a subspace of s. Thus, by Hahn-Banach theorem, l can extend to a continuous linear functional on s. By Theorem 4.15, there exists a sequence $\beta(\lambda)$ such that $l(\alpha(\lambda)) = \sum_{\lambda} \alpha(\lambda)\overline{\beta}(\lambda)$, where $\beta(\lambda)$ satisfies (4.19) with α replaced by β. By Theorem 4.13, $\sum_{\lambda} \beta(\lambda)\psi_\lambda(x)$ belongs to the space BMO. This ends the proof of the second part of Theorem 4.14.

We now consider the Hardy spaces $H^p(X)$, $p < 1$. The difference between $H^1(X)$ and $H^p(X)$ is that if $f \in H^p(X), p < 1$, then f is not necessarily a function in $L^p(X)$. More precisely, $H^p, p < 1$, is the collection of certain distributions. Following [CW2], we define the Hardy spaces $H^p, p < 1$, by the atomic decomposition method.

$H^p, p < 1$, is defined as the subspace of all $f \in (\dot{\mathcal{M}}_0(\beta,\gamma))'$ such that f can be written as

$$f(x) = \sum_{0}^{\infty} \lambda_k a_k(x) \tag{4.21}$$

where $\sum_{0}^{\infty} |\lambda_k|^p < \infty$ and a_k are $(p,2)$ atoms.

The $(p,2)$ atoms are defined as follows: for each atom $a_k(x)$, there exists a ball B_k, of volume $\mu(B_k)$, such that the support of $a_k(x)$ is contained in B_k,

$$\|a_k\|_2 \leq \mu(B_k)^{\frac{1}{2}-\frac{1}{p}}, \tag{4.22}$$

and

$$\int a_k(x) d\mu(x) = 0. \tag{4.23}$$

Under these conditions, the series (4.21) converges in the sense of distributions. More precisely, if $g(x) \in \dot{\mathcal{M}}_0(\beta,\gamma)$, the series $\sum_{0}^{\infty} \lambda_k \langle g, a_k\rangle$ is absolutely convergent, whose convergence is based on an obvious fact: there exists a constant C such that for each $g \in \dot{\mathcal{M}}_0(\beta,\gamma)$ and each $(p,2)-$atom $a(x)$ with the support B of radius r, then $|\langle g, a_k\rangle| \leq C\|g\|_{\dot{\mathcal{M}}(\beta,\gamma)}$.

To characterize the space $H^p, p < 1$, by use of wavelet expansions, we choose wavelets $\psi_\lambda, \lambda \in \Lambda$, with the regularity ϵ, and use all notations as in Theorem 4.12. We then have

Theorem 4.16. *Suppose that a distribution $f \in (\dot{\mathcal{M}}_0(\beta,\gamma))', 0 < \beta, \gamma < \theta$, and f has a wavelet series $f(x) = \sum_{\lambda\in\Lambda} \alpha(\lambda)\psi_\lambda(x)$. Then, for $\frac{1}{1+\epsilon} < p < 1$, the following ones are equivalent:*

$$f \in H^p(X); \tag{4.24}$$

$$\Big(\sum_{\lambda \in \Lambda} |\alpha(\lambda)|^2 |\psi_\lambda(x)|^2 \Big)^{\frac{1}{2}} \in L^p(X); \tag{4.25}$$

$$\Big(\sum_{\lambda \in \Lambda} |\alpha(\lambda)|^2 \mu(Q_{\lambda,\gamma})^{-1} \chi_{Q_{\lambda,\gamma}}(x) \Big)^{\frac{1}{2}} \in L^p(X); \tag{4.26}$$

and

$$\left(\sum_{\lambda \in \Lambda} |\alpha(\lambda)|^2 \mu(Q_\lambda)^{-1} \chi_{Q_\lambda}(x) \right)^{\frac{1}{2}} \in L^p(X). \tag{4.27}$$

The proof of the equivalence between (4.25) and (4.27) is a straightforward rewriting of that of Theorem 4.6. It depends on the fact that, for every sequence $\omega(\lambda), \lambda \in \Lambda$, of $\pm 1s$, the operator $T_\omega : L^2(X) \to L^2(X)$, defined by $T_\omega(\psi_\lambda) = \omega(\lambda)\psi_\lambda, \omega = (\omega(\lambda))_{\lambda \in \Lambda}$, extends to a continuous linear operator on H^p, $\frac{1}{1+\epsilon} < p < 1$. This result follows from a general result on H^p whose proof will be given in Section 4.6, see Theorem 4.27.

Proposition 4.17 *Let $T : L^2(X) \to L^2(X)$ be a Calderón-Zygmund operator. Suppose that $T^*(1) = 0$. Then T extends to a continuous linear operator on H^p, $\frac{1}{1+\theta} < p \leq 1$.*

The proof of the above Proposition will be given in Section 4.6.

The proofs of (4.26) $\iff$ (4.27) $\iff$ (4.24) are similar to the proofs of Theorem 4.12. We leave these details to the reader.

From [CW2] it is well known that the dual of $H^p(X)$, $\frac{1}{1+\theta} < p \leq 1$, is the homogeneous Hölder spaces $\dot{C}^\gamma$ with $\gamma = \frac{1}{p} - 1$. Furthermore, it is also well known that $\dot{C}^\gamma$ is equivalent to the so-called Campanato-Morrey spaces. These spaces are defined by a similar way to the space BMO.

Definition 4.18. We say that a locally integrable function $g(x)$ defined on X belongs to the Campanato-Morrey space $L(\beta, 2), \beta \geq 0$, if

$$\|g\|_{L(\beta,2)} = \sup_B \mu(B)^{-\beta} \left(\int_B |g(x) - \gamma_B|^2 \frac{d\mu(x)}{\mu(B)} \right)^{\frac{1}{2}}$$

where B are balls in X and γ_B is a constant depending only on B.

One should observe that $L(0,2) = BMO$ and 2 can be replaced by $q, 1 \leq q < \infty$.

In the following result, we will give a new characterization of $L(\beta, 2)$ with $\beta = \frac{1}{p} - 1$, in terms of the wavelet coefficients.

Theorem 4.19. *Let $g(x)$ be a function belonging to $L(\frac{1}{p} - 1, 2)$. Then its wavelet coefficients $\alpha(\lambda) = \langle g, \psi_\lambda \rangle$ satisfy generalized Carleson's condition: there exists a constant C such that, for each dyadic cube Q,*

$$\sum_{Q_\lambda \subset Q} |\alpha(\lambda)|^2 \le C(\mu(Q))^{\frac{2}{p}-1}. \tag{4.28}$$

Conversely, if the coefficients $\alpha(\lambda), \lambda \in \Lambda$, *satisfy (4.28), then the series* $\sum_\lambda \alpha(\lambda)\widetilde{\psi}_\lambda(x)$ *converges, in the* $\sigma(L(\frac{1}{p}-1,2), H^p)$*–topology, to a function of* $L(\frac{1}{p}-1,2)$.

The proof of this theorem is a straightforward rewriting of Theorem 4.13.

Using Theorem 4.19, we can give a proof of the duality between H^p and $L(\frac{1}{p}-1,2)$. Again, we choose the wavelets with the regularity ϵ.

Theorem 4.20. *Let* $g(x) \in L(\frac{1}{p}-1,2)$, $\frac{1}{1+\epsilon} < p < 1$. *Then* $g(x)$ *defines a continuous linear functional* ℓ *on* H^p *by*

$$\ell(f) = \sum_\lambda \alpha(\lambda)\overline{\beta(\lambda)}$$

where $\alpha(\lambda)$ *and* $\beta(\lambda)$ *are the wavelet coefficients of* f *and* g, *respectively.*

Conversely, every continuous linear functional on H^p, $\frac{1}{1+\epsilon} < p < 1$, *is defined in this way.*

The proof of this theorem is similar to the proof of Theorem 4.14. One only needs to define the sequence spaces s^p and c^p by $\|S(\alpha)(x)\|_p < \infty$ and $(\mu(Q))^{1-\frac{2}{p}} \sum_{Q_\lambda \subset Q} |\beta(\lambda)|^2 < \infty$, respectively. A similar proof to Theorem 4.15 implies that $(s^p)^* = c^p$. We leave these details to the reader.

4.6 Besov Spaces on Spaces of Homogeneous Type

We begin with recalling the definition of the Besov spaces on the standard case of $\mathbb{R}^n$. We shall define the homogeneous Besov spaces by use of the Littlewood -Paley decomposition $I = \sum_j \Delta_j$, where Δ_j is the operator of convolution with ψ_j as we described in the Introduction.

The homogeneous Besov space $\dot{B}_p^{s,q}(\mathbb{R}^n)$ when $s < \frac{n}{p}$ or $s = \frac{n}{p}$ and $q = 1$, is a space of distributions. This means that $\dot{B}_p^{s,q}(\mathbb{R}^n)$ is a vector-subspace of $\mathcal{S}'(\mathbb{R}^n)$ and that the inclusion $\dot{B}_p^{s,q}(\mathbb{R}^n) \subset \mathcal{S}'(\mathbb{R}^n)$ is continuous. A distribution $f \in \mathcal{S}'(\mathbb{R}^n)$ belongs to $\dot{B}_p^{s,q}(\mathbb{R}^n)$ if and only if the following two conditions are satisfied: (i) the partial sums $\sum_{-m}^{m} \Delta_j(f)$ converge to f for the $\sigma(\mathcal{S}', \mathcal{S})$–topology; (ii) $\|f\|_{\dot{B}_p^{s,q}} = \left\{ \sum_{j\in\mathbb{Z}} (2^{sj}\|\Delta_j(f)\|_p)^q \right\}^{\frac{1}{q}} < \infty$.

For example, when $s < \frac{n}{p}$, the function 1 does not belong to $\dot{B}_p^{s,q}(\mathbb{R}^n)$ because $\Delta_j(1) = 0$ for every $j \in \mathbb{Z}$.

If $s = \frac{n}{p}$ and $q > 1$, or if $s > \frac{n}{p}$, $\dot{B}_p^{s,q}(\mathbb{R}^n)$ is no long even a space of distributions, instead it is a quotient space. In these cases, the function 1 belongs to $\dot{B}_p^{s,q}(\mathbb{R}^n)$ but it is identified with the function which is identically zero.

Let $\sigma = s - \frac{n}{p} > 0$. The Besov space $\dot{B}_p^{s,q}(\mathbb{R}^n)$ can be defined as a subspace of the homogeneous Hölder space $\dot{C}^\sigma$. This inclusion is based on the fact that $\|\Delta_j(f)\|_\infty \le C2^{\frac{nj}{p}}\|\Delta_j(f)\|_p$. Therefore the series $\sum\limits_{-\infty}^{\infty}\Delta_j(f)$ converges to f in the quotient space. Moreover, the Besov spaces $\dot{B}_p^{s,q}(\mathbb{R}^n)$ are generalizations of the homogeneous Sobolev spaces $\dot{L}^{s,2}(\mathbb{R}^n) = \dot{B}_2^{s,2}(\mathbb{R}^n)$ and of the homogeneous Hölder spaces $\dot{C}^s(\mathbb{R}^n) = \dot{B}_\infty^{s,\infty}(\mathbb{R}^n)$.

In [HS] and [H2], the homogeneous Besov spaces on spaces of homogeneous type are similarly defined with convolution operators $\Delta_j(f)$ replaced by non -convolution operators $D_j(f)$, where $D_j(x,y)$, the kernel of D_j, are wavelets given in Theorem 3.15 and $f \in (\dot{\mathcal{M}}_0(\beta,\gamma))', 0 < \beta, \gamma < \theta$. The main result in this section is to give the following characterization of $\dot{B}_p^{s,q}(X)$ in terms of the wavelet coefficients.

Theorem 4.21. *Suppose that $f \in (\dot{\mathcal{M}}_0(\beta,\gamma))', 0 < \beta, \gamma < \theta$ and f has the wavelet expansion $\sum\limits_{\lambda}\alpha(\lambda)\widetilde{\psi}_\lambda(x)$. Then for $-\theta < s < \theta, \max\{\frac{1}{1+\theta}, \frac{1}{1+s+\theta}\} < p, q \le \infty, f \in \dot{B}_p^{s,q}(X)$ if and only if*

$$\left\{\sum_k\left(\sum_{\lambda\in\Lambda_k}(2^{k(s+\frac{1}{2}-\frac{1}{p})}|\alpha(\lambda)|)^p\right)^{\frac{q}{p}}\right\}^{\frac{1}{q}} < \infty$$

where $\Lambda_k = \{(k,\tau,\nu) : \tau \in I_k, 1 \le \nu \le N(k,\tau)\}$.

The proof of this theorem is based on the comparison property of the wavelet coefficients in Theorem 4.1. By choosing appropriate point from each $Q_\lambda, \lambda \in \Lambda_k$, we have

$$|D_k(f)(x)| \le (\mu(Q_\lambda))^{-\frac{1}{2}}|\langle f, \psi_\lambda\rangle| \le C2^{\frac{k}{2}}|\alpha(\lambda)|.$$

This implies one implication in Theorem 4.21. The other implication in Theorem 4.21 follows similarly.

The inhomogeneous Besov spaces can be studied by a similar way. We would like to leave these details to the reader.

4.7 The $T1$ Type Theorems

We have proved the $T1$ Theorem on $L^2(X)$ in Section 4 of Chapter 1. In this section we establish the $T1$ type theorems for spaces which are studied in the previous sections. To deal with certain applications, it is convenient to have several different ways of expressing the necessary and sufficient conditions which appear in the $T1$ type theorems.

We suppose, throughout this section, that the conditions (1.10), (1.11), (1.12) and Definition 1.14 of the weak boundedness property are all used for operators T and the corresponding kernels $K(x,y)$.

We have characterized some functional spaces in the previous sections by simple size conditions on the wavelet coefficients. Suppose that we are given a wavelet series $f(x) = \sum_{\lambda} \alpha(\lambda)\psi_\lambda(x)$ and an operator T. Formally, the wavelet series of $T(f)$ is given by

$$T(f)(x) = \sum_{\lambda}\sum_{\lambda'} \alpha(\lambda')\langle T(\psi_{\lambda'}), \widetilde{\psi}_\lambda\rangle \psi_\lambda(x).$$

Let us denote $\sigma(\lambda,\lambda') = \langle T(\psi_{\lambda'}), \widetilde{\psi}_\lambda\rangle$ and then we can consider $\sum_{\lambda'} \alpha(\lambda')$ $\sigma(\lambda,\lambda')$ as the wavelets coefficients of $T(f)$ with respect to the wavelets $\psi_\lambda, \lambda \in \Lambda$. To show that T is bounded on the certain space, by the characterization in terms of the wavelet coefficients, we only need to know the behavior of the wavelet coefficients of $T(f)$. This immediately follows from the properties of $\sigma(\lambda,\lambda')$. We will examine, in great depth, the relationship between all conditions (1.10)-(1.12), the weak boundedness property, $T(1)$ and $T^*(1)$. Roughly speaking, we will consider two kinds of conditions: the first kind of conditions is satisfied by $T(1) = T^*(1) = 0$, and the second ones are either $T(1) = 0$, or $T^*(1) = 0$. We will also describe the cases where $T(1) \neq 0$, or $T^*(1) \neq 0$.

We start with the following fundamental estimate on $\sigma(\lambda,\lambda')$.

Lemma 4.22. *Suppose that T is a Calderón-Zygmund singular integral operator with the kernel satisfying the conditions (1.10) and (1.11) (i.e. smoothness in the x variable only), and also $T(1) = 0, T \in WBP$ and that $\psi_\lambda, \lambda \in \Lambda = \{(k,\tau,\nu) : k \in \mathbb{Z}, \tau \in I_k, 1 \leq \nu \leq N(k,\tau)\}$, are wavelets used in Theorem 3.25. Then there exists a constant C independent of f such that*

$$|\sigma(\lambda,\lambda')| \leq C2^{-\frac{1}{2}(k'+k)}(2^{-(k-k')\epsilon'} \wedge 1)\frac{2^{-(k'\wedge k)\epsilon}}{(2^{-(k'\wedge k)} + \rho(y_\lambda, y_{\lambda'})^{1+\epsilon}} \tag{4.29}$$

where $0 < \epsilon' < \epsilon < \theta, y_{\lambda'}$ and y_λ are any fixed points in $Q_{\lambda'}$ and Q_λ, respectively.

By the construction of wavelets ψ_λ, we may assume that $\psi_\lambda(x), \lambda = (k,\tau,\nu)$, are supported in the ball $B(y_\lambda, 2^{-k})$. We first consider the case where $k' \geq k$ and $\rho(y_{\lambda'}, y_\lambda) \leq 4A^2 2^{-k}$. By the weak boundedness of T,

$$|\sigma(\lambda,\lambda')| \leq C2^{-k(1+2\eta)}\|\psi_{\lambda'}\|_\eta \|\psi_\lambda\|_\eta \leq C2^{-\frac{1}{2}(k+k')}2^k$$

which yields (4.29) for the case where $k' \geq k$ and $\rho(y_{\lambda'}, y_\lambda) \leq 4A^2 2^{-k}$.

Consider the case where $k' \geq k$ and $\rho(y_{\lambda'}, y_\lambda) > 4A^2 2^{-k}$. Recall that $\psi_\lambda(x) = \sqrt{\mu(Q_\lambda)} D_k(y_\lambda, x)$. By the fact that $D_k(1) = 0$, we can write

$$
\begin{aligned}
|\sigma(\lambda, \lambda')| &= \sqrt{\mu(Q_{\lambda'}} \sqrt{\mu(Q_\lambda)} \left| \iint D_k(y_\lambda, u) K(u, v) D_{k'}(y_{\lambda'}, v) d\mu(u) d\mu(v) \right| \\
&= \sqrt{\mu(Q_{\lambda'}} \sqrt{\mu(Q_\lambda)} \left| \iint D_k(y_\lambda, u)[K(u, v) - K(y_\lambda, v)] D_{k'}(y_{\lambda'}, v) d\mu(u) d\mu(v) \right| \\
&\leq C 2^{-\frac{1}{2}(k+k')} \iint \frac{2^{-k\epsilon}}{(2^{-k} + \rho(y_\lambda, u))^{1+\epsilon}} \frac{2^{-k\epsilon}}{\rho(y_\lambda, y_{\lambda'})^{1+\epsilon}} \frac{2^{-k'\epsilon}}{(2^{-k'} + \rho(v, y_{\lambda'}))^{1+\epsilon}} \\
&\quad \times d\mu(u) d\mu(v) \\
&\leq C 2^{-\frac{1}{2}(k+k')} 2^{-k\epsilon} \rho(y_\lambda, y_{\lambda'})^{-(1+\epsilon)}
\end{aligned}
$$

which implies (4.29) for the case where $k' \geq k$ and $\rho(y_\lambda, y_{\lambda'}) > 4A^2 2^{-k}$.

Now Consider the case where $k' < k$ and $\rho(y_\lambda, y_{\lambda'}) > 4A^2 2^{-k'}$. In this case, using the fact that $D_k(1) = 0$, we have

$$
\begin{aligned}
&|\sigma(\lambda, \lambda')| \\
&= \sqrt{\mu(Q_{\lambda'})} \sqrt{\mu(Q_\lambda)} \left| \iint D_k(y_\lambda, u)[K(u, v) - K(y_\lambda, v)] D_{k'}(y_{\lambda'}, v) d\mu(u) d\mu(v) \right| \\
&\leq C 2^{-\frac{1}{2}(k+k')} \iint \frac{2^{-k\epsilon}}{(2^{-k} + \rho(y_\lambda, u))^{1+\epsilon}} \frac{2^{-k\epsilon}}{\rho(y_\lambda, y_{\lambda'})^{1+\epsilon}} \frac{2^{-k'\epsilon}}{(2^{-k'} + \rho(v, y_{\lambda'})^{1+\epsilon}} \\
&\quad \times d\mu(u) d\mu(v) \\
&\leq C 2^{-\frac{1}{2}(k+k')} 2^{-k\epsilon} \rho(y_\lambda, y_{\lambda'})^{-(1+\epsilon)}
\end{aligned}
$$

which implies (4.29) for the case whenever $k' < k$ and $\rho(y_\lambda, y_{\lambda'}) > 4A^2 2^{-k'}$.

Finally, we consider the crucial case where $k' < k$ and $\rho(y_\lambda, y_{\lambda'}) \leq 4A^2 2^{-k'}$. Using the facts that $D_k(1) = T(1) = 0$, we get

$$
\begin{aligned}
&\iint D_k(y_\lambda, u) K(u, v)[D_{k'}(y_{\lambda'}, v) - D_{k'}(y_{\lambda'}, y_\lambda)] d\mu(u) d\mu(v) \\
&= \iint D_k(y_\lambda, u) K(u, v)[D_{k'}(y_{\lambda'}, v) - D_{k'}(y_{\lambda'}, y_\lambda)] \eta_1\Big(\frac{\rho(v, y_\lambda)}{4A^2 2^{-k}}\Big) d\mu(u) d\mu(v) \\
&\quad + \iint D_k(y_\lambda, u)[K(u, v) - K(y_\lambda, v)][D_{k'}(y_{\lambda'}, v) - D_{k'}(y_{\lambda'}, y_\lambda)] \eta_2\Big(\frac{\rho(v, y_\lambda)}{4A^2 2^{-k}}\Big) \\
&\quad \times d\mu(u) d\mu(v) \\
&= A + B,
\end{aligned}
$$

where $\eta_1 \in C^1(R), \eta_1(x) = 1$ for $|x| \leq 4$ and $\eta_1(x) = 0$ for $|x| \geq 6$, and $\eta_2 = 1 - \eta_1$.

With $\psi_\lambda(u) = \sqrt{\mu(Q_\lambda)} D_k(y_\lambda, u)$ and $\phi_{\lambda'}(v) = \sqrt{\mu(Q_{\lambda'})}[D_{k'}(y_{\lambda'}, v) - D_{k'}(y_{\lambda'}, y_\lambda)] \eta_1(\frac{\rho(v, y_\lambda)}{2^{-k}})$, then, by the fact that $T \in WBP$,

$$\begin{aligned}|\langle T\phi_{\lambda'},\psi_\lambda\rangle|| &= |\sqrt{\mu(Q_\lambda)}\sqrt{\mu(Q_{\lambda'})}A| \\ &\le C2^{-\frac{1}{2}(k'+k)}2^{-k(1+2\eta)}\|\phi_{\lambda'}\|_\eta\|\psi_\lambda\|_\eta \\ &\le C2^{-\frac{1}{2}(k'+k)}2^{-k(1+2\eta)}\{2^{-k(\epsilon+\eta)}2^{k'(1+\epsilon)}\}\{2^{k(1+\eta)}\} \\ &\le C2^{-\frac{1}{2}(k'+k)}2^{-k\epsilon}2^{k'(1+\epsilon)} = C2^{-\frac{1}{2}(k'+k)}2^{(k'-k)\epsilon}2^{k'}\end{aligned}$$

which is dominated by the right-hand side of (4.29) whenever $k' < k$ and $\rho(y_\lambda, y_{\lambda'}) \le 4A^2 2^{-k'}$.

Using the smoothness of $K(x,y)$ in x, together with

$$|D_{k'}(y_{\lambda'}, v) - D_{k'}(y_{\lambda'}, y_\lambda)| \le C2^{k'}$$

and

$$|D_{k'}(y_{\lambda'}, v) - D_{k'}(y_{\lambda'}, y_\lambda)| \le C2^{k'}\left(\frac{\rho(y_\lambda, v)}{2^{-k'} + \rho(y_\lambda, y_{\lambda'})}\right)^\epsilon$$

when $\rho(y_\lambda, v) \le cA2^{-k'}$, we then have

$$\begin{aligned}&|\sqrt{\mu(Q_\lambda)}\sqrt{\mu(Q_{\lambda'})}B| \\ &\le C2^{-\frac{1}{2}(k'+k)}2^{-k\epsilon}2^{k'}\int_{\rho(y_\lambda,v)\ge cA2^{-k'}}\rho(y_\lambda, v)^{-(1+\epsilon)}d\mu(v) \\ &\quad +C2^{-\frac{1}{2}(k'+k)}2^{(k'-k)\epsilon}\int_{cA2^{-k'}\ge\rho(y_\lambda,v)\ge cA2^{-k}}\rho(y_\lambda, v)^{-1}d\mu(v) \\ &\le C(1+\log 2^{(k-k')})2^{\frac{1}{2}(k'+k)}2^{(k'-k)\epsilon}2^{k'}\end{aligned}$$

which, again, is dominated by the right-hand side of (4.29) when $k' < k$ and $\rho(y_\lambda, y_{\lambda'}) \le 4A^2 2^{-k'}$. This completes the proof of Lemma 4.22.

If adding the conditions (1.12) and $T^*(1) = 0$ into Lemma 4.22 and repeating the similar proof, we have

Lemma 4.23. *Suppose that T is a Calderón-Zygmund singular integral operator with kernel satisfying the conditions (1.10), (1.11) and (1.12), and also $T(1) = T^*(1) = 0, T \in WBP$ and that $\psi_\lambda, \lambda \in \Lambda = \{(k,\tau,\nu) : k \in \mathbb{Z}, \tau \in I_k, 1 \le \nu \le N(k,\tau)\}$, are wavelets used in Theorem 3.25. Then there exists a constant C such that*

$$|\sigma(\lambda, \lambda')| \le C2^{-\frac{1}{2}(k'+k)}2^{-|k-k'|\epsilon}\frac{2^{-(k'\wedge k)\epsilon}}{(2^{-(k'\wedge k)} + \rho(y_\lambda, y_{\lambda'})^{1+\epsilon}}. \tag{4.30}$$

We are now on the position to prove the $T1$ type theorems. We start with the first kind of the $T1$ theorem.

Theorem 4.24. *Suppose that T is a Calderón-Zygmund singular integral operator with kernel satisfying the conditions (1.10), (1.11), (1.12), $T(1) = T^*(1) = 0$, and $T \in WBP$. Then T is bounded on all spaces studied in the previous sections, namely $L^p, 1 < p < \infty; \dot{L}^{s,p}, -\epsilon < s < \epsilon, \max\{\frac{1}{1+\epsilon}, \frac{1}{1+s+\epsilon}\} <$*

$p < \infty; \dot{C}^s, -\epsilon < s < \epsilon; H^p, \frac{1}{1+\epsilon} < p \le 1; BMO$ and $\dot{B}_p^{s,q}, -\epsilon < s < \epsilon, \max\{\frac{1}{1+\epsilon}, \frac{1}{1+s+\epsilon}\} < p, q \le \infty$. Moreover, the norms of Tf on these spaces are dominated by a constant times the norm of f on the corresponding spaces.

Note first that $L^p(X) = \dot{L}^{p,0}, 1 < p < \infty, \dot{L}^{p,0} = H^p, \frac{1}{1+\epsilon} < p \le 1, (H^1)^* = BMO$, and $\dot{B}_\infty^{s,\infty} = \dot{C}^s$ if $-\epsilon < s < \epsilon$. Thus, we only need to prove Theorem 4.24 for spaces $\dot{L}^{p,s}$ and $\dot{B}_p^{s,q}$.

Suppose that $f(x) = \sum\limits_\lambda \alpha(\lambda)\psi_\lambda(x)$. Then we can write $T(f)(x)$, formally, by

$$T(f)(x) = \sum_\lambda \sum_{\lambda'} \sigma(\lambda, \lambda')\alpha(\lambda')\widetilde{\psi}_\lambda(x).$$

Therefore, to show that T is bounded on $\dot{L}^{p,s}$, we only need to prove that there exists a constant C such that, for

$$\left\| \left\{ \sum_{\lambda\in\Lambda} \left[2^{k(s+\frac{1}{2})} |\beta(\lambda)| \chi_{Q_\lambda} \right]^2 \right\}^{\frac{1}{2}} \right\|_p \le C \left\| \left\{ \sum_{\lambda\in\Lambda} \left[2^{k(s+\frac{1}{2})} |\alpha(\lambda)| \chi_{Q_\lambda} \right]^2 \right\}^{\frac{1}{2}} \right\|_p$$

where $\beta(\lambda) = \sum\limits_{\lambda'} \sigma(\lambda, \lambda')\alpha(\lambda')$. To see this, by Lemma 4.23, we have

$$|\beta(\lambda)| \le C \sum_{k'} 2^{-\frac{1}{2}(k'+k)} 2^{|k-k'|\epsilon} \sum_{\lambda'\in\Lambda_{k'}} |\alpha(\lambda')| \frac{2^{-(k'\wedge k)\epsilon}}{(2^{-(k'\wedge k)} + \rho(y_\lambda, y_{\lambda'})^{1+\epsilon}}. \tag{4.31}$$

To estimate the last term above, we use the following lemma.

Lemma 4.25. *Suppose that $0 < \epsilon < \theta$ and $\frac{1}{1+\epsilon} < r < p$. Then there exists a constant C such that, for each $x \in Q_\lambda$,*

$$\sum_{\lambda'\in\Lambda_{k'}} |\alpha(\lambda')| \frac{2^{-(k'\wedge k)\epsilon}}{(2^{-(k'\wedge k)} + \rho(y_\lambda, y_{\lambda'})^{1+\epsilon}}$$
$$\le C 2^{(k'\wedge k) + \frac{(k'-k\wedge k')}{r}} \{M(\sum_{\lambda'\in\Lambda_{k'}} |\alpha(\lambda')|^r \chi_{Q_{\lambda'}})(x)\}^{\frac{1}{r}}.$$

The proof of this lemma is easy. To see this, for fixed y_λ we let $B_0 = \{Q_{\lambda'} : \rho(y_\lambda, y_{\lambda'}) \le 2^{-(k'\wedge k)}$ and $B_j = \{Q_{\lambda'} : 2^{j-1}2^{-(k'\wedge k)} < \rho(y_\lambda, y_{\lambda'}) \le 2^j 2^{-(k'\wedge k)}$ for $j \ge 1$. Then

$$\sum_{\lambda'\in\Lambda_{k'}} |\alpha(\lambda')| \frac{2^{-(k'\wedge k)\epsilon}}{(2^{-(k'\wedge k)} + \rho(y_\lambda, y_{\lambda'}))^{1+\epsilon}}$$
$$\le C 2^{(k'\wedge k)} \sum_{j\ge 0} 2^{-j(1+\epsilon)} \Bigg(\sum_{Q_{\lambda'}\in B_j} |\alpha(\lambda')|^r \Bigg)^{\frac{1}{r}}$$

$$\leq C\sum_{j\geq 0} 2^{(k'\wedge k)} 2^{\frac{(k'-k\wedge k')}{r}} 2^{-j((1+\epsilon)-\frac{1}{r})}\Big(\int \sum_{Q_{\lambda'}\in B_j} |\alpha(\lambda')|^r \chi_{Q_{\lambda'}}\Big)^{\frac{1}{r}}$$

$$\leq C 2^{(k'\wedge k)} 2^{\frac{(k'-k\wedge k')}{r}} \sum_{j\geq 0} 2^{-j((1+\epsilon)-\frac{1}{r})}\Big(M\Big(\sum_{Q_{\lambda'}\in B_j} |\alpha(\lambda')|\chi_{Q_{\lambda'}}\Big)^r(x)\Big)^{\frac{1}{r}}.$$

Summing over j yields the result.

Applying this lemma and Hölder inequality implies

$$\sum_{\lambda\in\Lambda} (2^{k(s+\frac{1}{2})}|\beta(\lambda)|)^2 \chi_{Q_\lambda}$$
$$\leq C\sum_{\lambda}\Big(\sum_{k'} 2^{ks}2^{-\frac{1}{2}k'}2^{-|k-k'|\epsilon}2^{(k'\wedge k)}2^{\frac{(k'-k\wedge k')}{r}}$$
$$\times \Big(M\Big(\sum_{\lambda'\in\Lambda_{k'}} |\alpha(\lambda')|\chi_{Q_{\lambda'}}\Big)^r\Big)^{\frac{1}{r}}\Big)^2 \chi_{Q_\lambda}$$
$$\leq C\sum_{\lambda}\Big(\sum_{k'} 2^{(k-k')s}2^{-k'}2^{-|k-k'|\epsilon}2^{(k'\wedge k)}2^{\frac{(k'-k\wedge k'k)}{r}}$$
$$\times \Big(M\Big(\sum_{\lambda'\in\Lambda_{k'}} 2^{k'(s+\frac{1}{2})}|\alpha(\lambda')|\chi_{Q_{\lambda'}}\Big)^r\Big)^{\frac{1}{r}}\Big)^2 \chi_{Q_\lambda}$$
$$\leq C\sum_{\lambda}\sum_{k'} 2^{(k-k')s}2^{-k'}2^{-|k-k'|\epsilon}2^{(k'\wedge k)}2^{\frac{(k'-k\wedge k'k)}{r}}$$
$$\times \Big(M\Big(\sum_{\lambda'\in\Lambda_{k'}} 2^{k'(s+\frac{1}{2})}|\alpha(\lambda')|\chi_{Q_{\lambda'}}\Big)^r\Big)^{\frac{2}{r}} \chi_{Q_\lambda}$$
$$\leq C(M(\sum_{\lambda'} 2^{k'(s+\frac{1}{2})}|\alpha(\lambda')|\chi_{Q_{\lambda'}})^r(x))^{\frac{2}{r}}$$

where the last inequality follows from the facts that if $\frac{1}{1+s+\epsilon} < r < p$ and $0 < s < \epsilon$, then

$$\sum_{k'} 2^{(k-k')s}2^{-k'}2^{-|k-k'|\epsilon}2^{(k'\wedge k)}2^{\frac{(k'-k\wedge k')}{r}} \leq C$$

and

$$\sum_{k} 2^{(k-k')s}2^{-k'}2^{-|k-k'|\epsilon}2^{(k'\wedge k)}2^{\frac{(k'-k\wedge k')}{r}} \leq C.$$

Taking square root and the L^p norm on the both side, and then applying the Fefferman-Stein vector valued maximal function inequality yield the desired result.

We now prove that T is bounded on $\dot{B}_p^{s,q}, \max\{\frac{1}{1+\epsilon}, \frac{1}{1+s+\epsilon}\} < p, q \leq \infty$. By the characterization of $\dot{B}_p^{s,q}$ in terms of the wavelet coefficients, we only need to show

$$\Big\{\sum_k \Big(\sum_{\lambda\in\Lambda_k} (2^{ks}2^{k(\frac{1}{2}-\frac{1}{p})}|\beta(\lambda)|)^p\Big)^{\frac{q}{p}}\Big\}^{\frac{1}{q}} \leq C\Big\{\sum_k \Big(\sum_{\lambda\in\Lambda_k} (2^{ks}2^{k(\frac{1}{2}-\frac{1}{p})}|\alpha(\lambda)|)^p\Big)^{\frac{q}{p}}\Big\}^{\frac{1}{q}}$$

where $\alpha(\lambda)$ and $\beta(\lambda)$ are same as above.

Using the same notation and the estimates in (4.33), we have

$$\begin{aligned}
&\sum_k \Big(\sum_{\lambda\in\Lambda_k} (2^{ks}2^{k(\frac{1}{2}-\frac{1}{p})}|\beta(\lambda)|)^p\Big)^{\frac{q}{p}} \\
&\leq C\sum_k \Big(\sum_{\lambda\in\Lambda_k} \Big(\sum_{\lambda'} 2^{ks}2^{k(\frac{1}{2}-\frac{1}{p})}2^{-\frac{1}{2}(k'+k)}2^{|k-k'|\epsilon} \\
&\qquad\times\frac{2^{-(k'\wedge k)\epsilon}}{2^{-(k'\wedge k)}+\rho(y_\lambda,y_{\lambda'})^{1+\epsilon}}|\alpha(\lambda')|\Big)^p\Big)^{\frac{q}{p}} \\
&= C\sum_k \Big(\sum_{\lambda\in\Lambda_k} \Big(\sum_{k'}\sum_{\lambda'\in\Lambda_{k'}} 2^{(k-k')s}2^{|k-k'|\epsilon}\frac{2^{k'}2^{-(k'\wedge k)\epsilon}}{(2^{-(k'\wedge k)}+\rho(y_\lambda,y_{\lambda'}))^{1+\epsilon}} \\
&\qquad\times[2^{k'(s+\frac{1}{2}-\frac{1}{p})}|\alpha(\lambda')|]\Big)^p\Big)^{\frac{q}{p}}.
\end{aligned}$$

Applying Hölder inequality for $p > 1$ and the $p-$inequality $(a+b)^p \leq a^p + b^p, a, b > 0$ for $\frac{1}{1+\epsilon} < p \leq 1$, and then taking the sum over $\lambda \in \Lambda_k$, imply that the last term above is dominated by

$$\begin{aligned}
&C\sum_k \Big(\sum_{k'} 2^{(k-k')s(p\wedge 1)}2^{|k-k'|\epsilon(p\wedge 1)}2^{-(k\wedge k')(1-(p\wedge 1))}2^{k'(1-(p\wedge 1))} \\
&\qquad\times\sum_{\lambda'\in\Lambda_{k'}} [2^{k'(s+\frac{1}{2}-\frac{1}{p})}|\alpha(\lambda')|]^p\Big)^{\frac{q}{p}}.
\end{aligned}$$

Finally, applying Hölder inequality for $\frac{q}{p} > 1$, and the $\frac{q}{p}-$inequality for $\frac{q}{p} \leq 1$, and then interchanging the summations over k and k' yield

$$\sum_k \Big(\sum_{\lambda\in\Lambda_k} (2^{ks}2^{k(\frac{1}{2}-\frac{1}{p})}|\beta(\lambda)|)^p\Big)^{\frac{q}{p}} \leq C\Big(\sum_{k'}\Big(\sum_{\lambda'\in\Lambda'} 2^{k'(s+\frac{1}{2}-\frac{1}{p})}|\alpha(\lambda')|\Big)^p\Big)^{\frac{q}{p}}.$$

Taking the $\frac{1}{q}$th power gives the desired result.

One should observe that Theorem 4.24 gives the another proof of the $T1$ theorem. More precisely, the proof of Theorem 4.24 implies the first step of the proof of the $T1$ theorem when $T(1) = T^*(1) = 0$.

If the only half conditions are satisfied, then we have the following second kind of the $T1$ theorem ([L] and [DH]).

Theorem 4.26. *Suppose that T is a Calderón-Zygmund singular integral operator with kernel satisfying the conditions (1.10), (1.11), $T(1) = 0$, and $T \in WBP$. Then T is bounded on $\dot{L}^{s,p}, 0 < s < \epsilon, \frac{1}{1+\epsilon} < p < \infty$ and $\dot{B}_p^{s,q}, 0 < s < \epsilon, \frac{1}{1+\epsilon} < p, q \leq \infty$. Moreover, the norm of Tf on these spaces are dominated by a constant times the norm of f on the corresponding spaces.*

Replacing Lemma (4.23) by Lemma (4.22) in the proof of Theorem 4.24 gives the proof of Theorem 4.26. We leave details to the reader.

In Theorem 4.24 and Theorem 4.26, the conditions $T(1) = 0$, or $T^*(1) = 0$ are sufficient conditions. A natural problem is that when these conditions are also necessary. The following result answers this problem.

Theorem 4.27. *Suppose that T is a Calderón-Zygmund operator. Then T extends to a continuous linear operator on $H^p, \frac{1}{1+\epsilon} < p \leq 1$, if and only if $T^*(1) = 0$.*

The condition $T^*(1) = 0$ is clearly necessary. Indeed, $\int T(\psi)(x)d\mu(x) = 0$ for all wavelets $\psi(x)$. The integral is well defined since $T(\psi)(x) = O((\rho(x, x_0))^{-1-\epsilon}$ as $\rho(x, x_0) \to \infty$ for any fixed $x_0 \in X$. By the definition, this means $T^*(1) = 0$.

To show that the condition $T^*(1) = 0$ is also sufficient, as G. Weiss remarked in [CW2], it is enough to prove that T maps each atom of H^p to molecule of H^p. However, we would like to give another proof which uses Theorem 4.24 and a result about the para-product operator.

We now define the para-product operator Π_b. Let ϕ_λ be father functions and ψ_λ be wavelets constructed by Coifman's idea as given in Theorem 3.25. Let b be a function in BMO. We define the para-product operator Π_b by

$$\Pi_b(f)(x) = \sum_\lambda \alpha(\lambda)\beta(\lambda)\widetilde{\psi}_\lambda(x)$$

where $\alpha(\lambda) = \langle b, \psi_\lambda \rangle, \beta(\lambda) = \langle f, \phi_\lambda \rangle$ and $\widetilde{\psi}_\lambda$ are given by Theorem 3.25.

It is easy to see that $\Pi_b(1) = b, (\Pi_b)^*(1) = 0$ and the kernel of Π_b satisfies all conditions (1.10), (1.11) and (1.12). Moreover, Π_b is bounded on $L^2(X)$. To see this, we apply Π_b to an arbitrary function $f \in L^2(X)$. By Theorem 4.6, we have

$$\|\Pi_b(f)\|_2^2 \leq C \sum_\lambda |\alpha(\lambda)|^2 |\beta(\lambda)|^2.$$

To find an upper bound for this last sum, we use Carleson's well-known lemma.

Lemma 4.28. *Let $p(\lambda), \lambda \in \Lambda$, be a sequence of positive numbers such that $\sum_{Q_\lambda \subset Q} p(\lambda) \leq \mu(Q)$, for every dyadic cube Q. Then, for every sequence $\omega(\lambda) \geq 0, \lambda \in \Lambda$, we have*

$$\sum_{\lambda \in \lambda} \omega(\lambda) p(\lambda) \leq \int_X \omega(x) d\mu(x)$$

where $\omega(x) = \sup_{x \in Q_\lambda} \omega(\lambda)$.

See [MC] for the proof of this lemma. In our case, if we ignore the constants, we have $p(\lambda) = |\alpha(\lambda)|^2, \omega(\lambda) = |\beta(\lambda)|^2$ and $\omega(x) = (M(f))^2(x)$, where $M(f)$ is the Hardy-Littlewood maximal function of f. We finish by observing that $\int_X (M(f))^2(x) d\mu(x) \leq C\|f\|_2^2$. We return to the proof of the second part of Theorem 4.27. It is now enough to show that Π_b is bounded on $H^p, \frac{1}{1+\epsilon} < p \leq 1$ because, by Theorem 4.24, $\widetilde{T} = T - \Pi_{T(1)}$ is bounded on H^p since $\widetilde{T}(1) = (\widetilde{T})^*(1) = 0$. We will use atomic decomposition to show that Π_b is bounded on H^p.

Moreover, as G. Weiss remarked in [CW2], it suffices to check that $\|\Pi_b(a)\|_{H^p} \leq C$, where a is any H^p-atom and the constant C is independent of a. To this end, we may assume that a is an atom with the support $Q \in \Lambda_0$. Let $\widetilde{Q} \in \Lambda_{-8}$ and $4Q \subset \widetilde{Q}$. We denote $\beta(\lambda) = \langle a, \phi_\lambda \rangle$. By the characterization of H^p in terms of the wavelet coefficients, one only needs to show that there exists a constant C such that $\|S(a)(x)\|_p \leq C$, where

$$S(a)(x) = \{\sum_\lambda |\alpha(\lambda)|^2 |\beta(\lambda)|^2 (\mu(Q_\lambda))^{-1} \chi_{Q_\lambda}(x)\}^{\frac{1}{2}}.$$

Since Π_b is bounded on $L^2(X)$, by the conditions on a, we have

$$\int_{\widetilde{Q}} S^p(x) d\mu(x) \leq (\mu(\widetilde{Q}))^{1-\frac{p}{2}} \|a\|_2^p \leq C.$$

Note that if $x \notin \widetilde{Q}$ and $\beta(\lambda) \neq 0$, then there is only one $Q_\lambda, \lambda \in \Lambda_k$ with $k \leq -4$. Moreover,

$$\beta(\lambda) = \int_Q a(y) \phi_\lambda(y) d\mu(y) = \int_Q a(y)[\phi_\lambda(y) - \phi_\lambda(y_0)] d\mu(y)$$

where y_0 is the center of Q.

By the size and smoothness conditions on ϕ_λ, the support condition on a and the fact that $|\alpha(\lambda)| \leq C 2^{-\frac{k}{2}}$, we get

$$|\alpha(\lambda)||\beta(\lambda)| \chi_{Q_\lambda}(x) \leq C 2^{-\frac{k}{2}} \frac{2^{-k\epsilon}}{(2^{-k} + \rho(x, y_0))^{1+2\epsilon}}$$

where $\lambda \in \Lambda_k, k \leq -4$.

This implies that if $x \notin \widetilde{Q}$, then $S(a)(x) \leq C \frac{1}{(\rho(x,y_0))^{1+\epsilon}}$ and hence

$$\int\limits_{(\widetilde{Q})^c} S^p(a)(x) d\mu(x) \leq C.$$

This gives the proof of Theorem 4.27.

Up to now, we have considered the continuity of Calderón-Zygmund singular integral operators T which satisfy the conditions $T(1) = 0$ or $T^*(1) = 0$. In what follows, we study the continuity of Calderón-Zygmund operators T which do not satisfy $T(1) = 0$. More precisely, we will consider the continuity of operators on $\dot{B}_1^{0,1}$. See [MM] for a similar question on the standard case of $\mathbb{R}^n$. Let us denote $\sigma(\lambda, \lambda') = \langle T\psi_{\lambda'}, \psi_\lambda \rangle$, where $\psi_\lambda, \lambda \in \Lambda$ are wavelets given as in Theorem 3.25. The following result gives a simple characterization of the matrices $M = \sigma(\lambda, \lambda'), (\lambda, \lambda') \in \Lambda \times \Lambda'$, corresponding to the continuous linear operators $T : \dot{B}_1^{0,1} \to \dot{B}_1^{0,1}$.

Proposition 4.29 *The operator T is continuous on $\dot{B}_1^{0,1}$ if and only if there exists a constant C such that, for all $\lambda' \in \Lambda'$,*

$$\sum_{\lambda \in \Lambda} |\sigma(\lambda, \lambda')| (\mu(Q_\lambda))^{\frac{1}{2}} \leq C (\mu(Q_{\lambda'})^{\frac{1}{2}}. \tag{4.32}$$

The proof of the above proposition is easy, because, by the characterization of $\dot{B}_1^{0,1}$ in terms of the wavelet coefficients, Theorem 4.21 and (4.32) mean that, for each $\lambda' \in \Lambda', T((\mu(Q'_\lambda))^{-\frac{1}{2}} \psi_{\lambda'})$ belongs to $\dot{B}_1^{0,1}$ with the norm bounded by a constant which is independent of $\psi_{\lambda'}$. This implies that

$$\begin{aligned} \|T(f)\|_{\dot{B}_1^{0,1}} &\leq \sum_\lambda |\alpha(\lambda)| (\mu(Q_\lambda))^{\frac{1}{2}} \|T((\mu(Q_\lambda))^{-\frac{1}{2}} \psi_\lambda)\|_{\dot{B}_1^{0,1}} \\ &\leq C \sum_\lambda |\alpha(\lambda)| (\mu(Q_\lambda))^{\frac{1}{2}} \leq C \|f\|_{\dot{B}_1^{0,1}}. \end{aligned}$$

The proof of the other implication is immediate because, again by the characterization of $\dot{B}_1^{0,1}$, if T is bounded on $\dot{B}_1^{0,1}$, then $T((\mu(Q_{\lambda'}))^{\frac{-1}{2}} \psi_{\lambda'})$ belongs to $\dot{B}_1^{0,1}$ and $\|T((\mu(Q_{\lambda'}))^{\frac{-1}{2}} \psi_{\lambda'})\|_{\dot{B}_1^{0,1}} \leq C$. This is equivalent to (4.32).

To characterize the boundedness of T on $\dot{B}_1^{0,1}$ in terms of the conditions on $T(1)$ and $T^*(1)$, we define the following generalized Carleson measure.

Definition 4.30. Suppose that $\alpha(\lambda)$ are wavelet coefficients of f corresponding to the wavelets $\psi_\lambda, \lambda \in \Lambda$. We say that $\alpha(\lambda)$ satisfies the generalized Carleson measure condition if there exists a constant C, for each dyadic cube Q, such that

$$\sum_{Q_\lambda \subset Q} |\alpha(\lambda)| (\mu(Q_\lambda))^{\frac{1}{2}} \leq C \mu(Q). \tag{4.33}$$

The following result provides an example that $T(1) = 0$ is not the necessary condition for the continuity of operators.

Theorem 4.31. *Suppose that T is a Calderón-Zygmund singular integral operator. Then T extends to a continuous linear operator on $\dot{B}_1^{0,1}$ if and only if $T^*(1) = 0$ and the wavelet coefficients of $T(1)$ satisfy the generalized Carleson measure condition (4.33).*

The condition $T^*(1) = 0$ is clearly necessary for T being continuous on $\dot{B}_1^{0,1}$ because $\dot{B}_1^{0,1} \subset H^1$. We first prove the "if" part. To do this, as in the proof of Theorem 4.27, it suffices to show that there exists a constant C such that

$$\|\Pi_b(2^{\frac{k'}{2}}\psi_{\lambda'})\|_{\dot{B}_1^{0,1}} \leq C$$

where Π_b is the para-product operator used in the proof of Theorem 4.27 and $b = T(1)$.

Because $\Pi_b(2^{\frac{k'}{2}}\psi'_\lambda)(x) = \sum_\lambda \alpha(\lambda)\beta(\lambda)\widetilde{\psi}_\lambda(x)$, where $\alpha(\lambda)$ are wavelet coefficients of $T(1)$ and $\beta(\lambda) = \langle 2^{\frac{k'}{2}}\psi_{\lambda'}, \phi_\lambda\rangle$, one only needs to show

$$\sum_\lambda |\alpha(\lambda)||\beta(\lambda)|(\mu(Q_\lambda))^{\frac{1}{2}} \leq C$$

where C is a constant. For each $\psi_{\lambda'}$, by the construction, its support is contained in $Q_{\lambda'}$. We split all dyadic cubes Q_λ into two groups: $G_1 = \{Q_\lambda : Q_\lambda \subseteq Q_{\lambda'}\}$ and $G_2 = \{Q_\lambda : Q_{\lambda'} \subset Q_\lambda\}$. Based on the conditions on the supports of $\psi_{\lambda'}$ and ϕ_λ, if $k' > k$ and $\beta(\lambda) \neq 0$, then there exists one and only one Q_λ such that $Q_{\lambda'} \subset Q_\lambda$. Suppose that $Q_{\lambda'} \subset Q_\lambda$ where $k' > k$ and $\beta(\lambda) \neq 0$. Then, by the estimate in Lemma 3.7,

$$|\beta(\lambda)| \leq C2^{-|k'-k|\epsilon}2^k$$

which together with the fact that $|\alpha(\lambda)| \leq C(\mu(Q_\lambda))^{\frac{1}{2}}$, implies

$$\sum_{Q_\lambda \in G_2} |\alpha(\lambda)||\beta(\lambda)|(\mu(Q_\lambda))^{\frac{1}{2}} \leq C.$$

If $Q_\lambda \subseteq Q_{\lambda'}$, the size conditions on $\psi_{\lambda'}$ and ϕ_λ imply that $|\beta| \leq C(\mu(Q_{\lambda'}))^{-1}$. Therefore, by (4.33),

$$\sum_{Q_\lambda \subset Q_{\lambda'}} |\alpha(\lambda)||\beta(\lambda)|(\mu(Q_\lambda))^{\frac{1}{2}} \leq C.$$

We now return to prove the "only if" part. It suffices to show that if the para-product operator Π_b where $b = T(1)$ is bounded on $\dot{B}_1^{0,1}$, then b satisfies the generalized Carleson measure condition (4.33). This is because if T is bounded on $\dot{B}_1^{0,1}$ then $T^*(1) = 0$. By the result above, $\widetilde{T} = T - \Pi_b$, is bounded on $\dot{B}_1^{0,1}$, where Π_b is the para-product operator with $b = T(1)$. Let us recall that $\{S_k(x,y)\}_{k\in\mathbb{Z}}$ are father functions constructed by Coifman's idea in Section 1.3 and $\{D_k(x,y)\}_{k\in\mathbb{Z}}$ are corresponding mother unctions. Let $\phi_\lambda = S_k(x,y_\lambda)$, for $\lambda \in \Lambda_k, \Lambda_k = \{(k,\tau,\nu) : \tau \in I_k, 1 \le \nu \le N(k,\tau)\}, \psi_\lambda(x) = \sqrt{Q}_\lambda D_k(y_\tau^{k,\nu}, x), \lambda \in \Lambda = \{(k,\tau,\nu) : k \in \mathbb{Z}, \tau \in I_k, 1 \le \nu \le N(k,\tau)\}$ be the wavelets given in Theorem 3.25. For each given dyadic cube P, let $100P$ be the dyadic cube containing P with 100 times side length as P. We denote ψ_{λ_0} by the wavelet corresponding to $100P$. Then it is not difficult to check that $\|\Pi_b(2^{\frac{k_0}{2}}\psi_{\lambda_0})\|_{\dot{B}_1^{0,1}} \le C$ and

$$\|\Pi_b(2^{\frac{k_0}{2}}\psi_{\lambda_0})\|_{\dot{B}_1^{0,1}} \approx \sum_\lambda |\alpha(\lambda)||\beta(\lambda)|(\mu(Q_\lambda))^{\frac{1}{2}}$$

where $\beta(\lambda) = \langle \frac{k_0}{2}\psi_{\lambda_0}, \phi_\lambda\rangle$.

One observe that $|\beta(\lambda)| = |\langle 2^{\frac{k_0}{2}}\psi_{\lambda_0}, \phi_\lambda\rangle| \ge C2^{k_0}$ whenever $Q_\lambda \subseteq P$, for some constant C independent of k_0. Therefore we have

$$\sum_{Q_\lambda\subseteq P} |\alpha(\lambda)|(\mu(Q_\lambda))^{\frac{1}{2}} \le C2^{-k_0}\sum_\lambda |\alpha(\lambda)||\beta(\lambda)|(\mu(Q_\lambda))^{\frac{1}{2}}$$
$$\le C2^{-k_0}\|\Pi_b(2^{\frac{k_0}{2}}\psi_{\lambda_0})\|_{\dot{B}_1^{0,1}} \le C2^{-k_0} \le C\mu(P)$$

which implies that $b = T(1)$ satisfies the generalized Carleson measure condition (4.33).

Before finishing this chapter, we give a nice application of the $T1$ theorem. By using the wavelet coefficients, we have given characterizations of all spaces which are studied in previous sections. However, by the Littlewood-Paley theory developed by G. David, J. L. Journé and S. Semmes, the $L^p, 1 < p < \infty$, spaces can be characterized in terms of an approximation to the identity without requiring the condition (3.19), namely the double Lipschitz condition. One should observe that the condition (3.19) plays a crucial role in developing the wavelet analysis in Chapter 3. Therefore, a natural question arises: If one can characterize all spaces studied in previous sections by use of an approximation to the identity with only conditions (3.14)-(3.18). As an application of the $T1$ theorem, we will give a positive answer about this question. The results given below demonstrate that the Calderón-Zygmund operator theory, namely the $T1$ theorems, are a powerful tool not only for the boundedness of operators but also for the study of spaces of functions and distributions.

Since the Hardy spaces H^p are special generalized Sobolev spaces $\dot{L}^{p,0}$ and the Hölder spaces C^s are special Besov spaces $\dot{B}_\infty^{s,\infty}$, in what follows, we only give new characterizations of $\dot{L}^{p,s}$ and $\dot{B}_p^{s,q}$.

Theorem 4.32. *Suppose that $\{P_k\}_{k\in\mathbb{Z}}$ is an approximation to the identity whose kernels satisfy only conditions (3.14)-(3.18). Let $Q_k = P_k - P_{k-1}$. Then for $f \in L^2(X)$ and $-\theta < s < \theta$,*

$$\|f\|_{\dot{B}_q^{s,p}} \approx \left\{ \sum_{k\in\mathbb{Z}} (2^{ks}\|Q_k(f)\|_p)^q \right\}^{\frac{1}{q}} \tag{4.34}$$

for $\max\left(\frac{1}{1+\theta}, \frac{1}{1+\theta+s}\right) < p \le \infty$ *and* $0 < q \le \infty$,

$$\|f\|_{\dot{L}^{s,p}} \approx \left\| \left\{ \sum_{k\in\mathbb{Z}} (2^{ks}|Q_k(f)|)^2 \right\}^{\frac{1}{2}} \right\|_p \tag{4.35}$$

for $\max\left(\frac{1}{1+\theta}, \frac{1}{1+\theta+s}\right) < p < \infty$.

We first prove that if $f \in L^2(X)$,

$$\left\{ \sum_{k\in\mathbb{Z}} (2^{ks}\|Q_k(f)\|_p)^q \right\}^{\frac{1}{q}} \le C\|f\|_{\dot{B}_q^{s,p}} \tag{4.36}$$

for $-\theta < s < \theta$ and $\max(\frac{1}{1+\theta}, \frac{1}{1+\theta+s}) < p \le \infty$ and $0 < q \le \infty$,

$$\left\| \left\{ \sum_{k\in\mathbb{Z}} (2^{ks}|Q_k(f)|)^q \right\}^{\frac{1}{q}} \right\|_p \le C\|f\|_{\dot{L}^{s,p}} \tag{4.37}$$

for $-\theta < s < \theta$ and $\max\left(\frac{1}{1+\theta}, \frac{1}{1+\theta+s}\right) < p, q < \infty$.

The $T1$ theorem does not play any role for these proofs in (4.36) and (4.37). Suppose that $f = \sum_{\lambda\in\Lambda} \alpha(\lambda)\widetilde{\psi}_\lambda(x)$ and $-\theta < s < \theta$ and $\max\left(\frac{1}{1+\theta}, \frac{1}{1+\theta+s}\right) < p \le \infty$ and $0 < q \le \infty$. By Theorem 3.25, we have

$$Q_j(f)(x) = \sum_{\lambda\in\Lambda} \alpha(\lambda)(Q_j\widetilde{\psi}_\lambda)(x).$$

By the almost orthogonal estimate in Lemma 3.7, $(Q_j\widetilde{\psi}_\lambda)(x)$ satisfies the following estimate

$$|(Q_j\widetilde{\psi}_\lambda)(x)| \le C2^{-\frac{1}{2}k}2^{-|j-k|\epsilon} \frac{2^{-(k\wedge j)\epsilon}}{(2^{-(k\wedge j)} + \rho(x, y_\tau^{k,\nu}))^{1+\epsilon}} \tag{4.38}$$

where $\lambda = (k, \tau, \nu)$ and $\epsilon < \theta$.

By first observing

$$\|Q_k(f)\|_p \le \sum_{\tau\in I_k} \sum_{\nu=1}^{N(k,\tau)} \left[(\mu(Q_\tau^{j,\nu}))^{\frac{1}{p}} \sup_{z\in Q_\tau^{k,\nu}} |Q_k(f)(z)| \right]^p \Bigg)^{\frac{1}{p}}$$

and then using the same proof of Theorem 4.1 and the estimate in (4.38), we obtain

$$\left\{\sum_{j\in\mathbb{Z}}(2^{js}\|Q_j(f)\|_p)^q\right\}^{\frac{1}{q}}$$

$$\leq\left\{\sum_{j\in\mathbb{Z}}\left(\sum_{\tau\in I_j}\sum_{\nu=1}^{N(j,\tau)}\left[(\mu(Q_\tau^{j,\upsilon}))^{-s+\frac{1}{p}}\sup_{z\in Q_\tau^{j,\upsilon}}|Q_j(f)(z)|\right]^p\right)^{\frac{q}{p}}\right\}^{\frac{1}{q}}$$

$$\leq C\left\{\sum_{k\in\mathbb{Z}}\left(\sum_{\lambda\in\Lambda_k}[(\mu(Q_\tau^{k,\nu}))^{-s+\frac{1}{p}+\frac{1}{2}}|a(\lambda)|]^p\right)^{\frac{q}{p}}\right\}^{\frac{1}{q}}\leq C\|f\|_{\dot{B}_q^{s,p}}$$

where the last inequality follows from the wavelet coefficients characterization for the Besov spaces in Theorem 4.24. This yields (4.36).

The proof of (4.37) is similar. We first write

$$\sum_{k\in\mathbb{Z}}(2^{ks}|Q_k(f)(x)|)^q=\sum_{k\in\mathbb{Z}}\sum_{\lambda\in\Lambda_k}2^{kqs}|Q_k(f)(x)|^q\chi_{Q_\lambda}(x).$$

The proof of (4.37) then follows from the estimate in (4.38) and the proof of Theorem 4.1.

The $T1$ theorem will be used in the proofs of the other implications of (4.36) and (4.37). More precisely, we decompose the identity operator on $L^2(X)$ by $I=R+S$, where

$$S(f)(x)=\sum_{k\in\mathbb{Z}}\sum_{\tau\in I_k}\sum_{\nu=1}^{N(k,\tau)}\mu(Q_\tau^{k,\nu})Q_k^N(x,y_\tau^{k,\upsilon})Q_k(f)(y_\tau^{k,\nu}).$$

Let us denote that $\sqrt{\mu(Q_\tau^{k,\nu})}Q_k^N(x,y_\tau^{k,\nu})=\widetilde{\theta}_\lambda(x)$ and $\sqrt{\mu(Q_\tau^{k,\nu})}Q_k(y_\tau^{k,\nu},x)=\theta_\lambda(x)$.

Then $S(f)$ defined above can be rewritten as

$$S(f)(x)=\sum_\lambda a(\lambda)\widetilde{\theta}_\lambda(x)$$

where $a(\lambda)=\langle f,\theta_\lambda\rangle$.

We shall show that S^{-1} is bounded in the norm of $\dot{B}_q^{s,p}$ and $\dot{L}^{s,p}$ for the range of s,p,q indicated in Theorem 4.32. Assuming, for the moment, that this has been done, by the wavelet coefficients characterization of the Besov spaces in Theorem 4.24 and Theorem 4.1, we have

$$\|S(f)\|_{\dot{B}_q^{s,p}}\leq C\left\{\sum_{k'\in\mathbb{Z}}\left(\sum_{\lambda'\in\Lambda_{k'}}[(\mu(Q_{\lambda'}))^{-s+\frac{1}{p}+\frac{1}{2}}|a(\lambda')|]^p\right)^{\frac{q}{p}}\right\}^{\frac{1}{q}}$$

$$\leq C\left\{\sum_{k\in\mathbb{Z}}(2^{ks}\|Q_k(f)\|_p)^q\right\}^{\frac{1}{q}}.$$

This, by the assumption that S^{-1} is bounded on $\dot{B}_q^{s,p}$, implies that

$$\|f\|_{\dot{B}_q^{s,p}} = \|S^{-1}S(f)\|_{\dot{B}_q^{s,p}} \leq C\Big\{\sum_{k\in\mathbb{Z}}(2^{ks}\|Q_k(f)\|_p)^q\Big\}^{\frac{1}{q}}.$$

Similarly, by the wavelet coefficients characterization of the generalized Sobolev spaces, Theorem 4.1 and the assumption that S^{-1} is bounded on $\dot{L}^{s,p}$, we have

$$\begin{aligned}\|f\|_{\dot{L}^{s,p}} = \|S^{-1}S(f)\|_{\dot{L}^{s,p}} &\leq C\Big\|\Big\{\sum_{\lambda'\in\Lambda'}[(\mu(Q_{\lambda'}))^{-s-\frac{1}{2}}|\alpha(\lambda')|\chi_{Q_{\lambda'}}(x)]^q\Big\}^{\frac{1}{q}}\Big\|_p \\ &\leq C\Big\|\Big\{\sum_{k\in\mathbb{Z}}(2^{ks}|Q_k(f)|)^q\Big\}^{\frac{1}{q}}\Big\|_p.\end{aligned}$$

Now all we need to do is to show that S^{-1} is bounded in the norm of $\dot{B}_q^{s,p}$ and $\dot{L}^{s,p}$ for the range of α, p, q indicated in Theorem 4.32.

As in the proof of Lemma 3.23, $R = I - S$, where $R = \widetilde{R}_N + R_N$ and $R_N(x,y)$, the kernel of R_N satisfies the conditions (3.35)-(3.37), (3.39) and $R_N(1) = R_N^*(1) = 0$, and the kernel of $\widetilde{R}_N$ satisfies the conditions (3.70)-(3.72), (3.74) and $\widetilde{R}_N(1) = \widetilde{R}_N^*(1) = 0$. So by the $T1$ Theorem 4.24, R is bounded on spaces $\dot{B}_q^{s,p}$ and $\dot{L}_2^{s,p}$ for the range of s, p, q indicated in Theorem 4.32 with norm less than $C2^{-N\delta} + C_N 2^{-j\delta}$. By choosing N large first and then j large enough, then S^{-1} is bounded on spaces $\dot{B}_q^{s,p}$ and $\dot{L}_2^{s,p}$ for the range of α, p, q indicated in Theorem 4.32.

Indeed, the Besov spaces of $\dot{B}_q^{s,p}$, for $0 < |s| < \theta$ and $1 \leq p, q \leq \infty$ and the generalized Sobolev spaces $\dot{L}^{s,p}$, for $0 < |s| < \epsilon$ and $1 < p, q < \infty$, can be characterized by more general operators whose kernels satisfy only half (depending on the sign of s) of the usual size, smoothness and cancellation conditions. More precisely, we have

Theorem 4.33. *Suppose that $\{P_k\}_{k\in\mathbb{Z}}$ is a sequence of operators whose kernels $P_k(x,y)$ satisfying the conditions 3.14, 3.15 and 3.17. Let $Q_k = P_k - P_{k-1}$. Then, for $f \in L^2$ and $0 < s < \theta$,*

$$\|f\|_{\dot{B}_q^{s,p}} \approx \Big\{\sum_{k\in\mathbb{Z}}(2^{ks}\|Q_k(f)\|_p)^q\Big\}^{\frac{1}{q}} \tag{4.39}$$

for $1 \leq p, q \leq \infty$, and for $1 < p, q < \infty$,

$$\|f\|_{\dot{L}^{s,p}} \approx \Big\|\Big\{\sum_{k\in\mathbb{Z}}(2^{ks}|Q_k(f)|)^q\Big\}^{\frac{1}{q}}\Big\|_p. \tag{4.40}$$

Indeed, repeating the same proof as in Theorem 4.32 and replacing the estimate in (4.38) by

$$|(Q_j\widetilde{\psi}_\lambda)(x)| \le C2^{-\frac{1}{2}k}(1\wedge 2^{-(j-k)\epsilon})\frac{2^{-(k\wedge j)\epsilon}}{(2^{-(k\wedge j)}+\rho(x,y_\tau^{k,\nu}))^{1+\epsilon}} \tag{4.41}$$

where $\lambda \in \Lambda_k$, we then obtain

$$\left\{\sum_{k\in\mathbb{Z}}(2^{ks}\|Q_k(f)\|_p)^q\right\}^{\frac{1}{q}} \le C\|f\|_{\dot{B}_q^{s,p}} \tag{4.42}$$

for $0 < s < \theta$ and $1 \le p, q \le \infty$, and

$$\left\|\left\{\sum_{k\in\mathbb{Z}}(2^{ks}|Q_k(f)|)^q\right\}^{\frac{1}{q}}\right\|_p \le C\|f\|_{\dot{L}^{s,p}} \tag{4.43}$$

for $0 < s < \theta$ and $1 < p, q < \infty$.

To show the other implications in (4.42) and (4.43), we need the following two lemmas.

Lemma 4.34. *Suppose $0 < s < \theta$. Then there exists a constant $C > 0$ such that*

$$\left\|\sum_{k-j>N} Q_jQ_k(f)\right\|_{\dot{B}_q^{s,p}} \le C2^{-Ns}\|f\|_{\dot{B}_q^{s,p}} \tag{4.44}$$

for $1 \le p, q \le \infty$, and

$$\left\|\sum_{k-j>N} Q_jQ_k(f)\right\|_{\dot{L}^{s,p}} \le C2^{-Ns}\|f\|_{\dot{L}^{s,p}} \tag{4.45}$$

for $1 < p, q < \infty$.

We first prove (4.44). By the definition of $\dot{B}_q^{s,p}$,

$$\begin{aligned}\left\|\sum_{k-j>N} Q_jQ_k(f)\right\|_{\dot{B}_q^{s,p}} &= \left\{\sum_{\ell\in\mathbb{Z}}\left(2^{\ell s}\left\|D_\ell\sum_{k-j>N}Q_jQ_k(f)\right\|_p\right)^q\right\}^{\frac{1}{q}} \\ &\le \left\{\sum_{\ell\in\mathbb{Z}}\left(2^{\ell s}\sum_{k-j>N}\|D_\ell Q_j\|_{p,p}\|Q_k(f)\|_p\right)^q\right\}^{\frac{1}{q}} \\ &\le C\left\{\sum_{\ell\in\mathbb{Z}}\left(2^{\ell s}\sum_{k-j>N}(2^{(j-\ell)\epsilon}\wedge 1)\|Q_k(f)\|_p\right)^q\right\}^{\frac{1}{q}}\end{aligned}$$

since $D_\ell Q_j(x,y)$ satisfies the estimate in Lemma 3.6. Therefore,

$$\begin{aligned}
&\Big\|\sum_{k-j>N} Q_jQ_k(f)\Big\|_{\dot{B}_q^{s,p}}\\
&= C\Big\{\sum_{\ell\in\mathbb{Z}}\Big(\sum_{k-j>N}(2^{(j-\ell)\epsilon}\wedge 1)2^{(\ell-j)s}2^{(j-k)s}2^{ks}\|Q_k(f)\|_p\Big)^q\Big\}^{\frac{1}{q}}\\
&\le C\Big\{\sum_{\ell\in\mathbb{Z}}\Big(\sum_{k-j>N}(2^{(j-\ell)\epsilon}\wedge 1)2^{(\ell-j)s}2^{(j-k)s}\Big)^{\frac{q}{q'}}\\
&\quad\times\Big[\sum_{k-j>N}(2^{(j-\ell)\epsilon}\wedge 1)2^{(\ell-j)s}2^{(j-k)s}(2^{ks}\|Q_k(f)\|_p)^q\Big]\Big\}^{\frac{1}{q}}\\
&\le C2^{-Ns}\|f\|_{\dot{B}_q^{s,p}},
\end{aligned}$$

where the last inequality follows from the facts that $0 < s < \epsilon < \theta$ and

$$\sum_{k-j>N}(2^{(j-\ell)\epsilon}\wedge 1)2^{(\ell-j)s}2^{(j-k)s} \le C2^{-Ns}. \tag{4.46}$$

The proof of (4.45) is similar. By the definition of $\dot{L}^{s,p}$,

$$\begin{aligned}
\Big\|\sum_{k-j>N} Q_jQ_k(f)\Big\|_{\dot{L}^{s,p}} &= \Big\|\Big\{\sum_{\ell\in\mathbb{Z}}\Big(2^{\ell s}\Big|D_\ell\sum_{k-j>N}Q_jQ_k(f)\Big|\Big)^q\Big\}^{\frac{1}{q}}\Big\|_p\\
&\le \Big\|\Big\{\sum_{\ell\in\mathbb{Z}}\Big(2^{\ell s}\sum_{k-j>N}|D_\ell Q_jQ_k(f)|\Big)^q\Big\}^{\frac{1}{q}}\Big\|_p\\
&\le C\Big\|\Big\{\sum_{\ell\in\mathbb{Z}}\Big(2^{\ell s}\sum_{k-j>N}(2^{(j-\ell)\epsilon'}\wedge 1)M(Q_k(f))\|_p\Big)^q\Big\}^{\frac{1}{q}}\Big\|_p
\end{aligned}$$

since $D_\ell Q_j(x,y)$ satisfies the estimate in Lemma 3.6. This yields

$$\begin{aligned}
&\Big\|\sum_{k-j>N} Q_jQ_k(f)\Big\|_{\dot{L}^{s,p}}\\
&= C\Big\|\Big\{\sum_{\ell\in\mathbb{Z}}\Big(\sum_{k-j>N}(2^{(j-\ell)\epsilon}\wedge 1)2^{(\ell-j)s}2^{(j-k)s}2^{ks}M(Q_k(f))\Big)^q\Big\}^{\frac{1}{q}}\Big\|_p\\
&\le C\Big\{\sum_{\ell\in\mathbb{Z}}\Big(\sum_{k-j>N}(2^{(j-\ell)\epsilon}\wedge 1)2^{(\ell-j)s}2^{(j-k)s}\Big)^{\frac{q}{q'}}\\
&\quad\times\Big[\sum_{k-j>N}(2^{(j-\ell)\epsilon}\wedge 1)2^{(\ell-j)s}2^{(j-k)s}(2^{ks}M(Q_k(f)))^q\Big]\Big\}^{\frac{1}{q}}\\
&\le C2^{-Ns}\|f\|_{\dot{L}^{s,p}}.
\end{aligned}$$

The last inequality follows from (4.46) and the Fefferman-Stein vector valued maximal inequality.

Lemma 4.35. *Suppose* $0 < s < \theta$. *Then there exist constants* $C > 0$ *and* $\delta > 0$ *such that*

$$\Big\| \sum_{j-k\geq N} Q_jQ_k(f)\Big\|_{\dot{B}_q^{s,p}} \leq C2^{-N\delta}\|f\|_{\dot{B}_q^{s,p}} \tag{4.47}$$

for $1 \leq p, q \leq \infty$, *and*

$$\Big\| \sum_{j-k\geq N} Q_jQ_k(f)\Big\|_{\dot{L}^{s,p}} \leq C2^{-N\delta}\|f\|_{\dot{L}^{s,p}} \tag{4.48}$$

for $1 < p, q < \infty$.

We write $\sum\limits_{j-k\geq N} Q_jQ_k(f) = \sum\limits_{j\geq N}\sum\limits_{k} Q_{j+k}Q_k(f)$ and denote $Q_k^j = Q_{j+k}Q_k$. We then have

$$|Q_k^j(x,y)| \leq C2^{-j\epsilon}\frac{2^{-k\epsilon}}{(2^{-k}+\rho(x,y))^{1+\epsilon}}, \tag{4.49}$$

$$|Q_k^j(x,y) - Q_k^j(x',y)| \leq C\Big(\frac{\rho(x,x')}{2^{-k}+\rho(x,y)}\Big)^{\epsilon}\frac{2^{-k\epsilon}}{(2^{-k}+\rho(x,y))^{1+\epsilon}} \tag{4.50}$$

for $\rho(x,x') \leq \frac{1}{2A}(2^{-k}+\rho(x,y))$, and

$$\int Q_k^j(x,y)d\mu(y) = 0. \tag{4.51}$$

The geometric mean between (4.49) and (4.50) implies that for $0 < \epsilon' < \epsilon$, and some $\delta > 0$,

$$|Q_k^j(x,y) - Q_k^j(x',y)| \leq C2^{-j\delta}\Big(\frac{\rho(x,x')}{2^{-k}+\rho(x,y)}\Big)^{\epsilon'}\frac{2^{-k\epsilon'}}{(2^{-k}+\rho(x,y))^{1+\epsilon'}} \tag{4.52}$$

for $\rho(x,x') \leq \frac{1}{2A}(2^{-k}+\rho(x,y))$.

Thus, by the definition of the Besov space and using Calderón's identity in Theorem 3.15,

$$\begin{aligned}
\Big\| \sum_{j-k\geq N} Q_jQ_k(f)\Big\|_{\dot{B}_q^{\alpha,p}} &= \Big\{\sum_{\ell\in\mathbb{Z}}\Big(2^{\ell s}\Big\|D_\ell\sum_{j>N}\sum_{k}Q_k^j\Big(\sum_i \widetilde{D}_iD_if\Big)\Big\|_p\Big)^q\Big\}^{\frac{1}{q}}\\
&= \Big\{\sum_{\ell\in\mathbb{Z}}\Big(\sum_{j>N}\sum_{k}\sum_{i}2^{\ell s}\|D_\ell Q_k^j\widetilde{D}_iD_if)\|_p\Big)^q\Big\}^{\frac{1}{q}}\\
&= \Big\{\sum_{\ell\in\mathbb{Z}}\Big(\sum_{j>N}\sum_{k\geq\ell}\sum_{i}2^{\ell s}\|D_\ell Q_k^j\widetilde{D}_iD_if)\|_p\Big)^q\Big\}^{\frac{1}{q}}\\
&\quad + \Big\{\sum_{\ell\in\mathbb{Z}}\Big(\sum_{j>N}\sum_{k<\ell}\sum_{i}2^{\ell s}\|D_\ell Q_k^j\widetilde{D}_iD_if)\|_p\Big)^q\Big\}^{\frac{1}{q}}\\
&= I + II.
\end{aligned}$$

Note that for $k < \ell$,

$$|D_\ell Q_k^j \widetilde{D}_i(x,y)| \le C2^{-j\delta}(2^{(k-\ell)\epsilon''} \wedge 1)(2^{(i-k)\epsilon''} \wedge 1)\frac{2^{-(k\wedge i\wedge \ell)\epsilon'}}{(2^{-(k\wedge i\wedge \ell)} + \rho(x,y))^{1+\epsilon'}}$$

where $0 < s < \epsilon'' < \epsilon$.

Therefore

$$\begin{aligned}
II &\le C\Bigg\{\sum_{\ell\in\mathbb{Z}}\Bigg(\sum_{j>N}\sum_{k<\ell}\sum_i 2^{\ell s}2^{-js}(2^{(k-\ell)\epsilon''} \wedge 1)(2^{(i-k)\epsilon''} \wedge 1)\|D_i f)\|_p\Bigg)^q\Bigg\}^{\frac{1}{q}}\\
&\le C2^{-N\delta}\Bigg\{\sum_{\ell\in\mathbb{Z}}\Bigg(\sum_{k<\ell}\sum_i 2^{\ell s}(2^{(k-\ell)\epsilon''} \wedge 1)(2^{(i-k)\epsilon''} \wedge 1)\|D_i f)\|_p\Bigg)^q\Bigg\}^{\frac{1}{q}}\\
&\le C2^{-N\delta}\Bigg\{\sum_{\ell\in\mathbb{Z}}\Bigg[\sum_{k<\ell}\sum_i 2^{(\ell-k)s}2^{(k-i)s}(2^{(k-\ell)\epsilon''} \wedge 1)(2^{(i-k)\epsilon''} \wedge 1)\Bigg]^{\frac{q}{q'}}\\
&\quad\times\Bigg[\sum_{k<\ell}\sum_i 2^{(\ell-k)s}2^{(k-i)s}(2^{(k-\ell)\epsilon''} \wedge 1)(2^{(i-k)\epsilon''} \wedge 1)(2^{is}\|D_i f)\|_p)^q\Bigg]\Bigg\}^{\frac{1}{q}}\\
&\le C2^{-N\delta}\Bigg\{\sum_i (2^{is}\|D_i f)\|_p)^q\Bigg\}^{\frac{1}{q}} \le C2^{-N\delta}\|f\|_{\dot{B}_q^{s,p}}.
\end{aligned}$$

Note that for $k \ge \ell$,

$$|D_\ell Q_k^j \widetilde{D}_i(x,y)| \le C2^{-j\delta}(2^{(i-k)\epsilon''} \wedge 1)\frac{2^{-(\wedge i\wedge \ell)\epsilon'}}{(2^{-(\wedge i\wedge \ell)} + \rho(x,y))^{1+\epsilon'}}$$

where $0 < s < \epsilon'' < \epsilon < \theta$.

Thus

$$\begin{aligned}
I &\le C\Bigg\{\sum_{\ell\in\mathbb{Z}}\Bigg(\sum_{j>N}\sum_{k\ge\ell}\sum_i 2^{\ell s}2^{-j\delta}(2^{(i-k)\epsilon''} \wedge 1)\|D_i f)\|_p\Bigg)^q\Bigg\}^{\frac{1}{q}}\\
&\le C2^{-N\delta}\Bigg\{\sum_{\ell\in\mathbb{Z}}\Bigg(\sum_{k\ge\ell}\sum_i 2^{\ell s}(2^{(i-k)\epsilon''} \wedge 1)\|D_i f)\|_p\Bigg)^q\Bigg\}^{\frac{1}{q}}\\
&\le C2^{-N\delta}\Bigg\{\sum_{\ell\in\mathbb{Z}}\Bigg[\sum_{k\ge\ell}\sum_i 2^{(\ell-k)s}2^{(k-i)s}(2^{(i-k)\epsilon''} \wedge 1)\Bigg]^{\frac{q}{q'}}\\
&\quad\times\Bigg[\sum_{k\ge\ell}\sum_i 2^{(\ell-k)s}2^{(k-i)s}(2^{(i-k)\epsilon''} \wedge 1)(2^{is}\|D_i f)\|_p)^q\Bigg]\Bigg\}^{\frac{1}{q}}\\
&\le C2^{-N\delta}\Bigg\{\sum_i (2^{is}\|D_i f)\|_p)^q]\Bigg\}^{\frac{1}{q}} \le C2^{-N\delta}\|f\|_{\dot{B}_q^{s,p}}.
\end{aligned}$$

These estimates imply (4.50). The proof of (4.51) is similar.

Now we return to the proof of Theorem 4.32.

We decompose the identity operator I by writing $I = R_N + T_N$, where $R_N = \sum_{|j-k|\geq N} Q_j Q_k$ and $T_N = \sum_k Q_k^N Q_k$ with $Q_k^N = \sum_{|j|\leq N} Q_{k+j}$. Lemma 4.33 and 4.34 imply that

$$\|R_N(f)\|_{\dot{B}_q^{s,p}} \leq C(2^{-Ns} + 2^{-N\delta})\|f\|_{\dot{B}_q^{s,p}} \tag{4.53}$$

for $0 < s < \epsilon$ and $1 \leq p, q \leq \infty$, and

$$\|R_N(f)\|_{\dot{L}^{s,p}} \leq C(2^{-Ns} + 2^{-N\delta})\|f\|_{\dot{L}^{s,p}} \tag{4.54}$$

for $0 < s < \epsilon$ and $1 < p, q < \infty$.

If we choose N large enough, then $(T_N)^{-1}$, the inverse of T_N, is bounded on $\dot{B}_q^{s,p}$, for $0 < s < \epsilon$ and $1 \leq p, q \leq \infty$, and $\dot{L}^{s,p}$, for $0 < s < \epsilon$ and $1 < p, q < \infty$. Now we have

$$\begin{aligned}
\|T_N(f)\|_{\dot{B}_q^{s,p}} &= \Big\{\sum_k (2^{ks}\|D_k T_N f\|_p)^q\Big\}^{\frac{1}{q}} \\
&\leq \Big\{\sum_k \Big(2^{ks}\Big\|D_k \sum_j Q_j^N Q_j f\Big\|_p\Big)^q\Big\}^{\frac{1}{q}} \\
&\leq \Big\{\sum_k \Big(2^{ks}\sum_j \|D_k Q_j^N\|_{p,p}\|Q_j f\|_p\Big)^q\Big\}^{\frac{1}{q}} \\
&\leq C\Big\{\sum_k \Big(\sum_j 2^{ks}(2^{(j-k)\epsilon'} \wedge 1)\|Q_j f\|_p\Big)^q\Big\}^{\frac{1}{q}} \\
&\leq C\Big\{\sum_j (2^{js}\|Q_j f\|_p)^q\Big\}^{\frac{1}{q}}.
\end{aligned}$$

This yields

$$\|f\|_{\dot{B}_q^{s,p}} \leq C\|T_N(f)\|_{\dot{B}_q^{s,p}} \leq C\Big\{\sum_j (2^{js}\|Q_j f\|_p)^q\Big\}^{\frac{1}{q}}.$$

Similarly,

$$\begin{aligned}
\|T_N(f)\|_{\dot{L}^{s,p}} &= \Big\|\Big\{\sum_k (2^{ks}|D_k T_N f|)^q\Big\}^{\frac{1}{q}}\Big\|_p \\
&\leq \Big\|\Big\{\sum_k (2^{ks}|D_k \sum_j Q_j^N Q_j f|)^q\Big\}^{\frac{1}{q}}\Big\|_p
\end{aligned}$$

$$\begin{aligned}
&\leq \left\| \left\{ \sum_k \left(2^{k\alpha} \sum_j |D_k Q_j^N Q_j f| \right)^q \right\}^{\frac{1}{q}} \right\|_p \\
&\leq C \left\| \left\{ \sum_k \left(\sum_j 2^{ks} (2^{(j-k)\epsilon'} \wedge 1) M(Q_j f)| \right)^q \right\}^{\frac{1}{q}} \right\|_p \\
&\leq C \left\| \left\{ \sum_j (2^{js} |Q_j f|)^q \right\}^{\frac{1}{q}} \right\|_p .
\end{aligned}$$

This implies

$$\|f\|_{\dot{L}^{s,p}} \leq C\|T_N(f)\|_{\dot{L}^{s,p}} \leq C \left\| \left\{ \sum_j (2^{js} |Q_j f|)^q \right\}^{\frac{1}{q}} \right\|_p$$

and the proof of Theorem 4.32 is completed.

Chapter 5
Littlewood-Paley Analysis on Non Homogeneous Spaces

5.1 Introduction

It is well know that the doubling property of the underlying measure is a basic hypothesis in the classical Calderón-Zygmund theory. A measure μ on $\mathbb{R}^n$ is said to be doubling if there exists some constant C such that $\mu(B(x,2r)) \leq c\mu(B(x,r))$ for all $x \in \operatorname{supp}(\mu), r > 0$, where $B(x,r) = \{y \in \mathbb{R}^n : |y-x| < r\}$. Recently it has been shown that many results of the classical Calderón-Zygmund theory also hold without assuming the doubling property. See [GM], [MMNO], [NTV1], [NTV2], [NTV3], [T1], [T2] and [T3] for more material.

Suppose that μ is a Radon measure on $\mathbb{R}^n$, which may be non-doubling and only satisfies the growth condition, namely there is a constant $C > 0$ such that for all $x \in \operatorname{supp}(\mu)$ and $r > 0$,

$$\mu(B(x,r)) \leq C_0 r^d \tag{5.1}$$

where $0 < d \leq n$.

One motivation for studying non doubling measures is the problem of analytic capacity which has a long history. More than a century ago Painlevé became interested in the problem of removable sets for bounded analytic functions. Let us consider a compact set E such that $\Omega = \mathbb{C} \setminus E$ is connected. Then E is a removable set for bounded analytic functions if any function f which is analytic and bounded in $\mathbb{C} \setminus E$ is a constant. In 1888 Painlevé proved that any compact set with zero one-dimensional Hausdorff measure is removable. In 1947 L. V. Ahlfors rephrased Painlevé's problem in terms of analytic capacity ([A]). The analytic capacity $\gamma(E)$ of E is

$$\gamma(E) = \sup\{\lim_{z\to\infty} |zf(z)| : f \in A(E)\}$$

with $A(E) = \{f : f \text{ analytic on } \Omega, \|f(z)\|_{L^\infty(\Omega)} \leq 1, \lim_{z\to\infty} f(z) = 0\}$. Then Ahlfors proved that E is removable for bounded analytic functions if and only if $\gamma(E) = 0$.

D. Deng and Y. Han, *Harmonic Analysis on Spaces of Homogeneous Type*, Lecture Notes in Mathematics 1966,

If the Hausdorff dimension of E is larger than 1, then the analytic capacity of E is positive. Thus Painlevé's problem is easy except for sets with Hausdorff dimension $\dim(E) = 1$ and Hausdorff measure $\Lambda_1(E) > 0$. For a subset E of a line, A. Denjoy proved that if $\Lambda_1(E) > 0$ then $\gamma(E) > 0$. Today we know that $\gamma(E) = \frac{1}{4}\Lambda_1(E)$ if $E \subset \mathbb{R}$. The well-known Denjoy conjecture says that $\Lambda_1(E) > 0 \Rightarrow \gamma(E) > 0$ whenever E is a subset of a rectifiable curve. The Denjoy conjecture was solved in 1977 by using the L^2 boundedness of the Cauchy integral on a Lipschitz curve with small Lipschitz constant, due to A. P. Calderón ([C2]). This was the first time that harmonic analysis entered the Painlevé problem. In 1967 Vitushkin conjectured that $\gamma(E) = 0$ if and only if E has zero Favard length, that is, the projection of E onto a line of almost every slope has zero length ([V]). However the Vitushkin conjecture on Favard length is false in general. P. W. Jones and T. Murai ([JM]) constructed a set E with zero Favard length but $\gamma(E) > 0$. Mattila, Melnikov and Verdera ([MMV]) in 1996 proved Vitushkin's Favard length conjecture in the case where E satisfies Ahlfors regularity

$$C^{-1}r \leq \Lambda_1(E \cap B(x,r)) \leq Cr$$

for all $z \in E$ and all $r < diam(E)$.

The one of main ideas in [MMV] is to use the Menger curvature and the formula

$$\int \left| \int \frac{d\mu(\zeta)}{z-\zeta} \right| d\mu(z) = \frac{c^2(\mu)}{6} + O\left(\int d\mu \right)$$

which is valid if μ satisfies the growth condition (5.1). This intriguing formula needs to be explained. See [MMV] for more details.

G. David proved Vitushkin's Favard length conjecture when E is not Ahlfors regular but $\Lambda_1(E) < \infty$. This is remarkable because the Ahlfors regularity of E implies that the measure $\mu(A) = \Lambda_1(A \cap E)$ is doubling ([D]). We remark that the doubling condition is necessary for the covering lemmas at the root of Calderón-Zygmund theory. A little later F. L. Nazarov, S. Treil and A. L. Volberg provided the $T(b)$ theorem on non homogeneous spaces which proves a conjecture of Vitushkin ([NTV3]).

Recently X. Tolsa gave a complete answer to the Painlevé problem ([T4]): A compact set $E \subset \mathbb{C}$ is non-removable for bounded analytic functions if and only if it supports a positive Radon measure μ with the growth condition (5.1) and finite curvature (the definition of the curvature of E with respect to μ will not be detailed here).

The purpose of this chapter is to give an outline of the Littlewood-Paley theory and functional spaces on non homogeneous spaces. Here $(\mathbb{R}^n, \mu)$ is called a non homogeneous space if the Radon measure μ defined on $\mathbb{R}^n$ satisfies the condition (5.1).

X. Tolsa in [T2] developed the Littlewood-Paley theory and used this theory to give another proof of the $T1$ theorem on non homogeneous spaces. We now describe Tolsa's results in the next section.

5.2 Littlewood-Paley Theory on Non Homogeneous Spaces

We will first describe Tolsa's construction that will derive the Littlewood-Paley theory on $L^2(\mu)$ for a measure μ satisfying (5.1) and non-doubling in general.

As in the case of spaces of homogeneous type, the key point is to construct a sequence of integral operators $\{S_k\}_{k\in\mathbb{Z}}$ given by kernels $S_k(x,y)$ defined on $\mathbb{R}^n \times \mathbb{R}^n$. This sequence of operators will yield some kind of approximation to the identity, namely $S_k \to I$ as $k \to \infty$ and $S_k \to 0$ as $k \to -\infty$ strongly in $L^2(\mu)$. The kernels $S_k(x,y)$ should satisfy some appropriate size and regularity conditions and

$$\int S_k(x,y)d\mu(y) = 1$$

for all $x \in \mathbb{R}^n$,

$$\int S_k(x,y)d\mu(x) = 1$$

for all $y \in \mathbb{R}^n$.

For each k we set $D_k = S_{k+1} - S_k$, and then, as in the case of spaces of homogeneous type, at least formally,

$$I = \sum_{k\in\mathbb{Z}} D_k.$$

Using Coifman's idea on the decomposition of the identity operator, we rewrite the above series as

$$\begin{aligned} I &= \Big(\sum_{k\in\mathbb{Z}} D_k\Big)\Big(\sum_{j\in\mathbb{Z}} D_j\Big) \\ &= \sum_{|k|\le N}\sum_{j\in\mathbb{Z}} D_{k+j}D_j + \sum_{|k|>N}\sum_{j\in\mathbb{Z}} D_{k+j}D_j = T_N + R_N. \end{aligned}$$

Repeating the same proof as in the case of spaces of homogeneous type, one can prove the Littlewood -Paley estimate on $L^2(\mu)$,

$$C^{-1}\|f\|_{L^2(\mu)} \le \sum_k \|D_k(f)\|^2_{L^2(\mu)} \le C\|f\|_{L^2(\mu)} \tag{5.2}$$

for all $f \in L^2(\mu)$.

Indeed, To prove the right inequality in (5.2) it suffices to show that the operator $\sum_k D_k^* D_k$ is bounded on $L^2(\mu)$, since $\sum_k \|D_k(f)\|^2_{L^2(\mu)} = \langle \sum_k D_k^* D_k(f), f\rangle$ and then Cotlar-Stein's lemma can be applied.

To obtain the left inequality in (5.2), under the appropriate conditions on D_k, we will see that $\|R_N\|_{2,2} < 1$ for N big enough. This implies that T_N is an invertible operator on $L^2(\mu)$, and hence $\|f\|_{L^2(\mu)} \leq C\|T_N(f)\|_{L^2(\mu)}$. This together with the right inequality in (5.2) implies the left inequality in (5.2). The $L^p, 1 < p < \infty$, estimates then follows from the theory of Calderón-Zygmund operators on non homogeneous spaces.

One of the difficulties for implementing the arguments above when μ is a non-doubling measure arises from the non trivial construction of the kernels $S_k(x, y)$ satisfying the required properties. When μ is a non-doubling measure the difficult step consists of obtaining functions $T_k(x, y)$ such that

$$\int T_k(x,y)d\mu(y) \approx 1$$

for each $x \in \text{supp}(\mu)$,

$$\int T_k(x,y)d\mu(x) \approx 1$$

for each $y \in \text{supp}(\mu)$.

As soon as the above functions $T_k(x, y)$ are constructed, then one can apply Coifman's idea to set $S_k = M_k T_k W_k T_k^* M_k$, with the same notation as in Chapter 2. Since $T_k(x, y) \neq T_k(y, x)$ in general, so $T_k \neq T^*$. To describe Tolsa's construction, for reader's convenience, we recall some basic notation and definitions, see [T2] and [T3] for more details.

Throughout this section, we denote by Q a closed cube with sides parallel to the axes, and by cQ the cube concentric with Q whose side length is c times the side length of Q.

We will assume that the constant C_0 in (5.1) has been chosen large enough so that for all cubes $Q \subset \mathbb{R}^n$ we have $\mu(Q) \leq C_0 \ell(Q)^d$, where $0 < d \leq n$ and $\ell(Q)$ is the side length of Q.

Definition 5.1. Given $\alpha > 1$ and $\beta > \alpha^d$, we say that the cube $Q \subset \mathbb{R}^n$ is (α, β)-doubling if $\mu(Q) \leq \beta\mu(Q)$.

Based on the growth condition (5.1), there are a lot of "big" and "small" doubling cubes. Given cubes $Q, R \subset \mathbb{R}^n$, we denote by z_Q the center of Q, and by Q_R the smallest cube concentric with Q containing Q and R.

Definition 5.2. Given two cubes $Q, R \subset \mathbb{R}^n$, we set

$$\delta(Q,R) = \max\left(\int_{Q_R \setminus Q} \frac{1}{|x-z_Q|^d} d\mu(x), \int_{R_Q \setminus R} \frac{1}{|x-z_R|^d} d\mu(x)\right).$$

Notice that $\ell(Q_R) \approx \ell(R_Q) \approx \ell(Q) + \ell(R) + dist(Q, R)$, and if $Q \subset R$, then $R_Q = R$ and $\ell(R) \leq \ell(Q_R) \leq 2\ell(R)$. We may treat points $x \in \text{supp}(\mu)$ as if they were cubes(with $\ell(x) = 0$). So for $x, y \in \text{supp}(\mu)$ and some cube Q, the notation $\delta(x, Q)$ and $\delta(x, y)$ make sense.

We now recall the definition of cubes of different generations. See again [T2] and [T3] for more details.

Definition 5.3. We say that $x \in \text{supp}(\mu)$ is a stopping point(or stopping cube) if $\delta(x, Q) < \infty$ for some cube $x \in Q$ with $0 < \ell(Q) < \infty$. We say that $\mathbb{R}^n$ is an initial cube if $\delta(Q, \mathbb{R}^n) < \infty$ for some cube Q with $0 < \ell(Q) < \infty$. The cubes Q such that $0 < \ell(Q) < \infty$ are called transit cubes.

It is easily seen that if $\delta(x, Q) < \infty$ for some transit cube Q containing x, the $\delta(x, Q') < \infty$ for any other transit cube Q' containing x. Also, if $\delta(Q, R^n) < \infty$ for some transit cube Q, then $\delta(Q', \mathbb{R}^n) < \infty$ for any transit cube Q'.

Definition 5.4. Assume that $\mathbb{R}^n$ is not an initial cube. We fix some doubling cube R_0. If Q is a transit cube, we say that Q is a cube of generation $k \geq 1$ if

$$\delta(Q, \mathbb{R}^n) = kA \pm \epsilon_1.$$

If $Q = x$ is a stopping cube, we say that Q is a cube of generation $k \geq 1$ if

$$\delta(x, \mathbb{R}^n) = kA \pm \epsilon_1.$$

Moreover, for all $k \leq 1$ we say that $\mathbb{R}^n$ is a cube of generation k.

In what follows, for any $x \in \text{supp}(\mu)$, we denote by $Q_{x,k}$ some fixed doubling cube centered at x of the kth generation. $\text{supp}(\mu)$ $\text{supp}(\mu)$
X. Tolsa in [T2] constructed functions $T_k(x, y)$ such that $\frac{1}{4} \leq \int T_k(x, y) d\mu$ (y), $\int T_k(x, y) d\mu(x) \leq \frac{3}{2}$. This enable him to define the operators S_k by $S_k = M_k T_k W_k T_k^* M_k$ where, as in the case of spaces of homogeneous type, M_k is the operator of multiplication by $m_k(x) = (T_k(1)(x))^{-1}$ and W_k is the operator of multiplication by $w_k(x) = (T_k^*(m_k)(x))^{-1}$. The following result gives an approximation to the identity on non homogeneous space.

Lemma 5.5. *There exists a sequence of operators* $\{S_k\}_{k \in \mathbb{Z}}$ *such that, for each* $k \in \mathbb{Z}, S_k(x, y)$*, the kernel of* S_k *satisfies the following properties:*

$$S_k(x, y) = S_k(y, x); \tag{5.3}$$

$$\int_{\mathbb{R}^n} S_k(x, y) d\mu(y) = 1, \qquad for\, x \in \text{supp}(\mu); \tag{5.4}$$

If $Q_{x,k}$ *is a transit cube, then*

$$\text{supp}(S_k(x, \cdot)) \subset Q_{x,k-1}; \tag{5.5}$$

If $Q_{x,k}$ *and* $Q_{y,k}$ *are transit cubes, then*

$$0 \leq S_k(x, y) \leq \frac{C}{(\ell(Q_{x,k}) + \ell(Q_{y,k}) + |x - y|)^d}; \tag{5.6}$$

If $Q_{x,k}, Q_{x',k}, Q_{y,k}$ *are transit cubes, and* $x, x' \in Q_{x_0,k}$ *for some* $x_0 \in \text{supp}(\mu)$*, then*

$$|S_k(x,y) - S_k(x',y)| \leq C \frac{|x-x'|}{\ell(Q_{x_0,k})} \frac{1}{(\ell(Q_{x,k}) + \ell(Q_{y,k}) + |x-y|)^d}; \quad (5.7)$$

If $Q_{x,k}, Q_{x',k}, Q_{y,k}, Q_{y',k}$ are transit cubes, $x, x' \in Q_{x_0,k}$ and $y, y' \in Q_{x_0,k}$ for some $x_0, y_0 \in \operatorname{supp}(\mu)$, then

$$\begin{aligned} &|[S_k(x,y) - S_k(x',y)] - [S_k(x,y') - S_k(x',y')]| \quad (5.8) \\ &\leq C \frac{|x-x'|}{\ell(Q_{x_0,k})} \frac{|y-y'|}{\ell(Q_{y_0,k})} \frac{1}{(\ell(Q_{x,k}) + \ell(Q_{y,k}) + |x-y|)^d}. \end{aligned}$$

See [T2] for the proof of (5.3)-(5.7) and [DHY] for the proof of (5.8).

For each $k \in \mathbb{Z}$, we set $D_k = S_k - S_{k-1}, E_k = \sum_{j \in \mathbb{Z}} D_{k+j} D_j$ and, for each $N \geq 1, \Phi_N = \sum_{|k| \leq N} E_k$. The following lemma gives the desired estimates which will be used to prove the Littlewood-Paley estimate on $L^2(\mathbb{R}^n, \mu)$.

Lemma 5.6. *We have*

(a) $\|D_j D_k\|_{2,2} \leq C 2^{-|j-k|\eta}$ *for all $j, k \in \mathbb{Z}$ and some $\eta > 0$;*

(b) $\sum_{k \in \mathbb{Z}} D_k = I$, *with strong convergence in $L^2(\mathbb{R}^n, \mu)$;*

(c) *The series $\sum_{j \in \mathbb{Z}} D_{k+j} D_j = E_k$ converges strongly in $L^2(\mathbb{R}^n, \mu)$ and*

$$\|E_k\|_{2,2} \leq C|k| 2^{-|k|\eta}$$

for all $k \in \mathbb{Z}$;

(d) $\Phi_N \to I$ *as $N \to \infty$ in the operator norm in $L^2(\mathbb{R}^n, \mu)$.*

By these estimates and an application of the Cotlar-Stein Lemma, namely Lemma 1.13, we get

Theorem 5.7. *If $f \in L^2(\mathbb{R}^n, \mu)$, then there exists a constant C such that*

$$C^{-1}\|f\|^2_{L^2(\mu)} \leq \sum_k \|D_k(f)\|^2_{L^2(\mu)} \leq C\|f\|^2_{L^2(\mu)}.$$

From this theorem we derive the following corollary.

Corollary 5.8. *Let $1 < p < \infty$. If $f \in L^p(\mathbb{R}^n, \mu)$, then there is a constant C such that*

$$C^{-1}\|f\|_{L^p(\mu)} \leq \left\| \Big(\sum_k \|D_k(f)\|^2 \Big)^{\frac{1}{2}} \right\|_{L^p(\mu)} \leq C\|f\|_{L^p(\mu)}. \quad (5.9)$$

The right inequality in (5.9) follows from the theory of vector valued Calderón-Zygmund operators. This theory has been extended by García-Cuerva and Martell [GM] to the case of non homogeneous spaces. The left inequality then follows from the right one by an argument similar to the one used for $p = 2$.

Using the Littlewood-Paley estimates, we will describe the proof of the $T1$ theorem on non homogeneous spaces in the next section.

5.3 The $T1$ Theorem on Non Homogeneous Spaces

In order to state the $T1$ theorem, we need to introduce some notation and definitions.

Definition 5.9. A kernel $K(x, y) : \mathbb{R}^n \times \mathbb{R}^n \setminus \{(x, y) : x = y\} \mapsto \mathbb{R}$ is called a (d-dimensional) Calderón-Zygmund(CZ) kernel if

(1) $|K(x, y)| \leq C|x - y|^{-d}$ if $x \neq y$;
(2) there exists $0 < \delta \leq 1$ such that

$$|K(x, y) - K(x', y)| + |K(y, x) - K(y, x')| \leq C \frac{|x - x'|^\delta}{|x - y|^{d+\delta}}$$

if $|x - x'| \leq \frac{1}{2}|x - y|$.

We say that T is a Calderón-Zygmund singular integral associated with the kernel $K(x, y)$ if for any compactly supported function $f \in L^2(\mathbb{R}^n, \mu)$

$$T(f)(x) = \int K(x, y) f(y) d\mu(y) \tag{5.10}$$

if $x \notin \text{supp}(f)$.

One should observe that the integral in (5.10) may be non convergent for $x \in \text{supp}(f)$, even for very nice functions. For this reason it is convenient to introduce the truncated operators $T_\epsilon, \epsilon > 0$:

$$T_\epsilon(f)(x) = \int_{|x-y|>\epsilon} K(x, y) f(y) d\mu(y).$$

It is now easy to see that this integral is absolutely convergent for any $f \in L^2(\mathbb{R}^n, \mu)$ and $x \in \mathbb{R}^n$.

We say that T is bounded on $L^2(\mathbb{R}^n, \mu)$ if the truncated operators T_ϵ are bounded on $L^2(\mathbb{R}^n, \mu)$ uniformly on $\epsilon > 0$.

Given a fixed constant $\kappa > 1$, we say that $f \in L^1_{loc}(\mathbb{R}^n, \mu)$ belongs to the space $BMO_\kappa(\mathbb{R}^n, \mu)$ if for some constant C

$$\sup_Q \frac{1}{\mu(\kappa Q)} \int_Q |f - m_Q(f)| d\mu \leq C$$

where the supremum is taken over all cubes Q and $m_Q(f)$ stands for the mean of f over Q with respect to μ, that is, $m_Q(f) = \frac{1}{\mu(Q)} \int\limits_Q f d\mu$.

As in the case of spaces of homogeneous type, we need weak boundedness property. This is the following definition.

Definition 5.10. We say that T is weakly bounded if

$$|\langle T_\epsilon \chi_Q, \chi_Q \rangle| \leq C\mu(Q) \tag{5.11}$$

for any cube Q, uniformly on $\epsilon > 0$.

Let us notice that this definition differs slightly from the usual definition of weak boundedness in spaces of homogeneous type. But, however, if T is bounded on $L^2(\mathbb{R}^n, \mu)$, then T is weakly bounded. See [NTV2] for more details.

Now we are in the position to state the $T1$ theorem.

Theorem 5.11. *If T is a Calderón-Zygmund singular integral operator which is weak bounded and $T_\epsilon(1), T_\epsilon^*(1) \in BMO_\kappa(\mathbb{R}^n, \mu)$ uniformly on $\epsilon > 0$ for some $\kappa > 1$, then T extends to a bounded operator on $L^2(\mathbb{R}^n, \mu)$.*

We remark that in the theorem, T_ϵ^* stands for the adjoint of T_ϵ with respect to the duality $\langle f, g \rangle = \int fg d\mu$. Moreover, T_ϵ and T_ϵ^* can be extended to $L^\infty(\mathbb{R}^n, \mu)$ in the usual way. Notice also that the kernel of the truncated operators T_ϵ do not satisfy the smoothness conditions in the definition of Calderón-Zygmund singular integral operator. For this reason one needs to introduce the regularized operators $\widetilde{T}_\epsilon$. To do this, let ϕ be a radial C^∞ function with $0 \leq \phi \leq 1$, vanishing on $B(0, \frac{1}{2})$ and identically equal to 1 on $\mathbb{R}^n \backslash B(0, 1)$. For each $\epsilon > 0$, we consider the integral operator $\widetilde{T}_\epsilon$ with the kernel $\phi(\frac{x-y}{\epsilon})K(x, y)$. It is not difficult to check that

$$|T_\epsilon(f) - \widetilde{T}_\epsilon(f)| \leq M_\mu(f)$$

where M_μ is the centered maximal Hardy-Littlewood operator with respect to measure $d\mu$. So T_ϵ is bounded on $L^2(\mathbb{R}^n, \mu)$ uniformly on $\epsilon > 0$ if and only if the same holds for $\widetilde{T}_\epsilon$. The kernel of $\widetilde{T}_\epsilon$ is L^∞ bounded and it is a CZ kernel with the constant uniformly on $\epsilon > 0$. One can also check that if $T_\epsilon(1), T_\epsilon^*(1) \in BMO_\kappa(\mathbb{R}^n, \mu)$ uniformly on $\epsilon > 0$ for some $\kappa > 1$, then these still hold for $\widetilde{T}_\epsilon$. The $T1$ theorem on non homogeneous spaces is proved first for the case where $T(1) = T^*(1) = 0$, and then the general case follows from using the para-product operators on non homogeneous spaces, as in the case of standard $\mathbb{R}^n$. See [T2] for more details.

We remark that the crucial ingredient used in [T2] is again the following identity:

$$I = T_N^{-1} T_N = \sum_{k \in \mathbb{Z}} T_N^{-1} D_k^N D_k = T_N T_N^{-1} = \sum_{k \in \mathbb{Z}} D_k D_k^N T_N^{-1}.$$

This identity is not a wavelet expanson, but, however, it is a start point to study the new functional spaces, which we describe now in the next section.

5.4 The Besov Space on Non Homogeneous Spaces

The purpose of this section is to study the Besov space on non homogeneous spaces. It is well known that the wavelet expansions(Calderón's identity) in Theorem 3.15 and Theorem 3.25 are crucial tools to study the Besov space on spaces of homogeneous type. Now the difficulty is that there are no such expansions. To be more precise, based on Tolsa's construction given in Lemma 5.5, we introduce the following definition of an approximation to the identity on non homogeneous spaces .

Definition 5.12. A sequence of operators $\{S_k\}_{k\in\mathbb{Z}})$ is said to be an approximation to the identity on $(\mathbb{R}^n, \mu)$ if $\{S_k(x,y)\}_{k\in\mathbb{Z}})$, the kernels of $\{S_k\}_{k\in\mathbb{Z}})$, satisfy the conditions (5.3)-(5.8).

Following [T2], we can write

$$I = T_N + R_N$$

where $T_N = \sum_{|k-j|\leq N} D_k D_j$ and $R_N = \sum_{|k-j|>N} D_k D_j$.

In [T2], as we describe in the previous section, X. Tolsa proved that T_N^{-1} exists and bounded on $L^2(\mathbb{R}^n, \mu)$. This immediately implies

$$I = \sum_{k\in\mathbb{Z}} D_k^N D_k T_N^{-1} \tag{5.12}$$

where $D_k^N = \sum_{|j|\leq N} D_{k+j}$ and the series converges in $L^2(\mathbb{R}^n, \mu)$.

The key observation to study the Besov space on non homogeneous spaces is the following result ([DHY]).

Theorem 5.13. *Let $\{S_k\}_{k\in\mathbb{Z}}, \{P_k\}_{k\in\mathbb{Z}}$ and $\{A_k\}_{k\in\mathbb{Z}}$ be approximations to the identity as defined in Definition 5.12. Set $E_k = P_k - P_{k-1}, G_k = A_k - A_{k-1}$ and R_N is defined by D_k as above. Then there exist constants C and $\nu, 0 < \nu < \frac{1}{2}$ such that, for all $1 \leq p, q \leq \infty, |s| < 2\nu\theta$, and for each $f \in L^2(\mathbb{R}^n, \mu)$,*

$$\left\{\sum_{j\in\mathbb{Z}} 2^{jsq} \|E_j R_N f\|_{L^p(\mu)}^q\right\}^{\frac{1}{q}} \tag{5.13}$$

$$\leq C(2^{-N(s+2\nu\theta)} + 2^{-N(2\nu\theta-s)})\left\{\sum_{k\in\mathbb{Z}} 2^{ksq} \|G_k f\|_{L^p(\mu)}^q\right\}^{\frac{1}{q}}.$$

Moreover if N is chosen such that $C(2^{-N(s+2\nu\theta)} + 2^{-N(2\nu\theta - s)}) < 1$, then, for each $f \in L^2(\mathbb{R}^n, \mu)$,

$$\left\{\sum_{j\in\mathbb{Z}} 2^{jsq}\|E_j T_N^{-1} f\|_{L^p(\mu)}^q\right\}^{\frac{1}{q}} \le C\left\{\sum_{k\in\mathbb{Z}} 2^{ksq}\|G_k f\|_{L^p(\mu)}^q\right\}^{\frac{1}{q}}. \qquad (5.14)$$

As a consequence of the above theorem, we have

Proposition 5.14 *Let $\{S_k\}_{k\in\mathbb{Z}}$ and $\{P_k\}_{k\in\mathbb{Z}}$ be approximations to the identity as defined in Definition 5.12. Set $D_k = S_k - S_{k-1}$ and $E_k = P_k - P_{k-1}$. Then, for all $1 \le p, q \le \infty$, $|s| < \theta$, and for each $f \in L^2(\mathbb{R}^n, \mu)$,*

$$\left\{\sum_{k\in\mathbb{Z}} 2^{ksq}\|D_k f\|_{L^p(\mu)}^q\right\}^{\frac{1}{q}} \approx \left\{\sum_{k\in\mathbb{Z}} 2^{ksq}\|E_k f\|_{L^p(\mu)}^q\right\}^{\frac{1}{q}}. \qquad (5.15)$$

The above proposition follows immediately from (5.12) and Theorem 5.13. Indeed, for each $f \in L^2(\mathbb{R}^n, \mu)$, by (5.12)

$$\left\{\sum_{j\in\mathbb{Z}} 2^{jsq}\|D_k f\|_{L^p(\mu)}^q\right\}^{\frac{1}{q}} \le \left\{\sum_{j\in\mathbb{Z}} 2^{jsq}\sum_{k\in\mathbb{Z}} \|E_j D_k^N D_k T_N^{-1} f\|_{L^p(\mu)}]^q\right\}^{\frac{1}{q}}.$$

By the estimate $\|E_j D_k^N\|_{2,2} \le C2^{-2\theta|j-k|}$ and applying Hölder inequality, we have

$$\left\{\sum_{j\in\mathbb{Z}} 2^{jsq}\|D_j f\|_{L^p(\mu)}^q\right\}^{\frac{1}{q}} \le C\left\{\sum_{k\in\mathbb{Z}} 2^{ksq}\|D_k T_N^{-1} f\|_{L^p(\mu)}^q\right\}^{\frac{1}{q}}.$$

Finally, the desired result follows from Theorem 5.13.

Proposition 5.14 leads to introduce the following subspaces of $L^2(\mathbb{R}^n, \mu)$.

Definition 5.15. Let $\{S_k\}_{k\in\mathbb{Z}}$ be an approximation to the identity as defined in Definition 5.12. Set $D_k = S_k - S_{k-1}$ for $k \in \mathbb{Z}$. Let θ be the regularity of the approximation to the identity $\{S_k\}_k$. For $|s| < \theta, 1 \le p, q \le \infty$, and $f \in L^2(\mathbb{R}^n, \mu)$, we define

$$\|f\|_{\dot{B}_q^{s,q}(\mu)} = \left\{\sum_{k\in\mathbb{Z}} 2^{ksq}\|D_k f\|_{L^p(\mu)}^q\right\}^{\frac{1}{q}}$$

and define a subspace of $L^2(\mathbb{R}^n, \mu)$ by

$$\dot{\mathcal{B}}_p^{s,q}(\mu) = \left\{f \in L^2(\mathbb{R}^n, \mu) : \|f\|_{\dot{B}_q^{s,q}(\mu)} < \infty\right\}.$$

Proposition 5.14 means that the definition of $\dot{\mathcal{B}}_p^{s,q}(\mu)$ is independent of the choice of the approximations to the identity. The following theorem is one of main results for the study of the Besov space on non homogeneous spaces ([DHY]).

Theorem 5.16. *Let θ and D_k be the same as in Definition 5.15. If $1 \leq p, q \leq \infty$ and for all $f \in \dot{\mathcal{B}}_p^{s,q}(\mu)$,*

$$f = \sum_{k \in \mathbb{Z}} D_k^N D_k T_N^{-1}(f) = \sum_{k \in \mathbb{Z}} T_N^{-1} D_k^N D_k(f) \tag{5.16}$$

holds in the norm of $\|\cdot\|_{\dot{B}_p^{s,q}(\mu)}$. Moreover, for all $g \in \dot{\mathcal{B}}_p^{s,q}(\mu)$ with $1 \leq p, q \leq \infty$,

$$\langle f, g\rangle = \sum_{k \in \mathbb{Z}} \langle D_k D_k^N T_N^{-1}(f), g\rangle = \sum_{k \in \mathbb{Z}} \langle T_N^{-1} D_k D_k^N(f), g\rangle \tag{5.17}$$

holds for $f \in \left(\dot{\mathcal{B}}_p^{s,q}(\mu)\right)'$, the dual of $\dot{\mathcal{B}}_p^{s,q}(\mu)$, with $1 \leq p, q \leq \infty$.

We remark that the condition (5.8) is crucial in the proof in (5.17). This theorem is not saying that we have the wavelet expansions on non homogeneous spaces in the usual way, because the present of the operator T_N^{-1}. But, however, the expansions in Theorem 5.16 can be considered as a generalized wavelet expansion since, by Theorem 5.13, T_N is bounded on $\dot{\mathcal{B}}_p^{s,q}(\mu)$ and hence the expansions in (5.16) provide a kind of wavelet expansion on non homogeneous spaces.

Now we are ready to introduce the Besov space on non homogeneous spaces.

Definition 5.17. Let θ and D_k be the same as in Definition 5.15. Let $|s| < \theta, 1 \leq p, q \leq \infty$, and p', q' be the conjugate index of p, q, respectively. We define

$$\dot{B}_p^{s,q}(\mu) = \left\{f \in (\dot{\mathcal{B}}_{p'}^{-s,q'}(\mu))' : \|f\|_{\dot{B}_p^{s,q}}(\mu) < \infty\right\}.$$

We can use the expansions in (5.16) to show that this definition does not depend on the choice of the approximations to the identity. Furthermore, we also can study the properties of the Besov space $\dot{B}_p^{s,q}(\mu)$, such that the dual space and the boundedness of operators on such a space, which include the Riesz operators defined via the approximation to the identity. See [DHY] for more details.

We remark that even if we obtain a kind of expansions as given in Theorem 5.16, we do not know if there is a discrete version of the wavelet expansions as given in Theorem 3.25 on spaces of homogeneous type.

References

[A] L. V. Ahlfors, Bounded analytic functions, Duke Math. J., **14** (1947), 1-11.

[C1] A. P. Calderón, Intermediate spaces and interpolation, the complex method, Studia Math. **24** (1964), 113-190.

[C2] A. P. Calderón, Proc. Cauchy integrals on Lipschitz curves and related operators, Nat. Acad. Sci. U.S.A. **74** (1977), 1324-1327.

[Ch1] M. Christ, Lectures on singular integral operators, CBMS Regional Conference Series in mathematics, Number 77. Published for the Conference Board of the Mathematical Sciences, Washington, D. C; by the American Mathematical Society, Providence, RI, (1990). x+132 pp.

[Ch2] M. Christ, A $T(b)$ theorem with remarks on analytic capacity and the Cauchy integral, Colloq. Math. **60/61** (1990), 601-628.

[CW1] R. Coifman and G. Weiss, Analyse harmornique non-commutative sur certains espaces homogenes. Lecture Notes in Math. Vol. 242. Springer-Verlag, Berlin-New York, (1971).

[CW2] R. Coifman and G. Weiss, Extensions of Hardy spaces and their use in analysis, Bull. Amer. Math. Soc. **83** (1977), 569-645.

[Da] I. Daubechies, The wavelet transform, time-frequency localization and signal analysis, IEEE Trans. Inf. Th., **36** (1990), 961-1105.

[D] G. David, Unrectifiable 1-sets have vanishing analytic capacity, Rev. Mat. Iberoamericana, **14** (1998), 369-479.

[DJ] G. David and J. L. Journé, A boundednees criterion for generalized Calderón-Zygmund operators, Ann. Math., **120** (1984), 371-397.

[DJS] G. David, J.-L. Journé, and S. Semmes, Opérateurs de Calderón-Zygmund, fonctions para-accrétives et interpolation, Rev. Mat. Iberoamericana **1** (1985), 1-56.

[DH] D. Deng and Y. S. Han, $T1$ theorem for Besov and Triebel-Lizorkin spaces, Sci. China Ser. A **48** (2005), 657-665.

[DHY] D. Deng, Y. S. Han and D. Yang, Besov spaces with non-doubling measures, Trans. Amer. Math. Soc., **358** (2005), 2965-3001.

[FJ] M. Frazier and B. Jawerth, A discrete transform and decompositions of distribution spaces, J. Func. Anal. **93** (1990), 34-170.

[FS] C. Fefferman and E. Stein, H^p spaces of several variables, Acta Math., **129** (1972), 137-193.

[GM] J. García-Cuerva and J. M. Martell, Weighted inequalities and vector-valued Calderón-Zygmund operators on non-homogeneous spaces, Publ. Mat., **44** (2000), 613-640.

[GrMo] A. Grossmann and J. Morlet, Decomposition of Hardy functions into square integrable wavelets of constant shape, SIAM J. Math. Anal., **15** (1984), 723-736.

[H1] Y. S. Han, Calderón-type reproducing formula and the Tb theorem, Rev. Mat. Iberoamericana **10** (1994), 51-91.

[H2] Y. S. Han, Plancherel-Pôlya type inequality on spaces of homogeneous type and its applications, Proc. Amer. Math. Soc. **126** (1998), 3315-3327.

[H3] Y. S. Han, Discrete Calderón type reproducing formula, Acta Math. Sin. (Engl. Ser.) **16** (2000), 277-294.

[H4] Y. S. Han, Inhomogeneous Calderón reproducing formula on spaces of homogeneous type, J. Geom. Anal. **7** (1997), 259-284.

[HS] Y. S. Han and E. Sawyer, Littlewood-Paley theory on spaces of homogeneous type and the classical function spaces, Mem. Amer. Math. Soc. **110** (1994), no 530, vi+126 pp.

[JM] P. W. Jones and T. Murai, Positive analytic capacity but zero Buffon needle probality, Pacific J. math., **133** (1988), 99-114.

[L] P. G. Lemarié, Continuité sur les espaces de Besov des opérateurs définis par des intégrales singulières, Ann. Inst. Fourier **35** (1985), 175-187.

[MMNO] J. Mateu, P. Mattila, A. Nicolau and J. Orobitg, BMO for nondoubling measures, Duke Math. J., **102** (2000), 533-565.

[MMV] P. Mattila, M. S. Melnikov and J. Verdera, The Cauchy integral, analytic capacity, and uniform rectifiability, Ann. of Math., **144** (1996), 127-136.

[MM] M. Meyer, Une classe d'espaces de type BMO. Applications aux intégrales singulières, Arkiv for Mat., **27** (1989), 305-318.

[M1] Y. Meyer, Les nouveaux opérateurs de Calderón-Zygmund, Astérisque **131** (1985), 237-254.

[M2] Y. Meyer, Wavelets and Operators, Translated by David Salinger, Cambridge Studies in Advanced mathematics, 37. Cambridge University Press, Cambridge, (1992), 224 pp.

[MC] Y. Meyer and R. Coifman, Wavelets Calderón-Zygmund and multilinear operators, Translated by David Salinger, Cambridge University Press, (1997), 315 pp.

[MS] R. Macias and C. Segovia, Lipschitz functions on spaces of homogeneous type, Adv. in Math. **33** (1979), 257-270.

[NTV1] F. Nararov, S. Treil and A. Volberg, Cauchy integral and Calderón -Zygmund operators on nonhomogeneous spaces, Internat. Math. Res. Notices **15** (1997), 703-726.

[NTV2] F. Nararov, S. Treil and A. Volberg, Weak type estimates and Cotlar inequalities for Calderón-Zygmund operators on nonhomogeneous spaces, Internat. Math. Res. Notices **9** (1998), 463-487.

[NTV3] F. Nararov, S. Treil and A. Volberg, The $Tb-$theorem on non homogeneous spaces, Acta Math., **190** (2003), 151-239.

[P] J. Peetre, On convolution operators leaving $L^{p,\lambda}$ spaces invariant, Ann. math. Pure Appl., **72** (1966), 295-304.

[NSW] A. Nagel, E. M. Stein, and S. Wainger, Balls and metrics defined by vector fields I: basic properties, Acta Math. **155** (1985), 103-147.

[SW] E. Stein and G. Weiss, On the theory of H^p spaces, Acta Math., **103**(1960), 25-62.

[T1] X. Tolsa, BMO, H^1, and Calderón-Zygmund operators for non doubling measures, Math. Ann., **319** (2001), 89-149.

[T2] X. Tolsa, Littlewood-Paley theory and the $T(1)$ theorem with non-doubling measures, Adv. in Math., **164** (2001), 57-116.

[T3] X. Tolsa, The space H^1 for nondoubling measures in terms of a grand maximal operator, Trans. Amer. Math. Soc., **355**(2003), 315-348.

[T4] X. Tolsa, Painlevé's problem and the semiadditivity of analytic capacity, Acta Math., **190** (2003), 105-149.

[V] A. G. Viyushkin, The analytic capacity of sets in problems of approximation theory, Uspekhi Mat. Nauk, **6** (1967), 141-199(Russian); English translation in Russian math. Surveys, **22**(1967), 139-200.

[Z] A. Zygmund, Trignometric Series, Second edition, Cambridge University Press, Cambridge, 1968.

Index

Lecture Notes in Mathematics

For information about earlier volumes
please contact your bookseller or Springer
LNM Online archive: springerlink.com

Vol. 1782: C.-H. Chu, A. T.-M. Lau, Harmonic Functions on Groups and Fourier Algebras (2002)
Vol. 1783: L. Grüne, Asymptotic Behavior of Dynamical and Control Systems under Perturbation and Discretization (2002)
Vol. 1784: L. H. Eliasson, S. B. Kuksin, S. Marmi, J.-C. Yoccoz, Dynamical Systems and Small Divisors. Cetraro, Italy 1998. Editors: S. Marmi, J.-C. Yoccoz (2002)
Vol. 1785: J. Arias de Reyna, Pointwise Convergence of Fourier Series (2002)
Vol. 1786: S. D. Cutkosky, Monomialization of Morphisms from 3-Folds to Surfaces (2002)
Vol. 1787: S. Caenepeel, G. Militaru, S. Zhu, Frobenius and Separable Functors for Generalized Module Categories and Nonlinear Equations (2002)
Vol. 1788: A. Vasil'ev, Moduli of Families of Curves for Conformal and Quasiconformal Mappings (2002)
Vol. 1789: Y. Sommerhäuser, Yetter-Drinfel'd Hopf algebras over groups of prime order (2002)
Vol. 1790: X. Zhan, Matrix Inequalities (2002)
Vol. 1791: M. Knebusch, D. Zhang, Manis Valuations and Prüfer Extensions I: A new Chapter in Commutative Algebra (2002)
Vol. 1792: D. D. Ang, R. Gorenflo, V. K. Le, D. D. Trong, Moment Theory and Some Inverse Problems in Potential Theory and Heat Conduction (2002)
Vol. 1793: J. Cortés Monforte, Geometric, Control and Numerical Aspects of Nonholonomic Systems (2002)
Vol. 1794: N. Pytheas Fogg, Substitution in Dynamics, Arithmetics and Combinatorics. Editors: V. Berthé, S. Ferenczi, C. Mauduit, A. Siegel (2002)
Vol. 1795: H. Li, Filtered-Graded Transfer in Using Noncommutative Gröbner Bases (2002)
Vol. 1796: J.M. Melenk, hp-Finite Element Methods for Singular Perturbations (2002)
Vol. 1797: B. Schmidt, Characters and Cyclotomic Fields in Finite Geometry (2002)
Vol. 1798: W.M. Oliva, Geometric Mechanics (2002)
Vol. 1799: H. Pajot, Analytic Capacity, Rectifiability, Menger Curvature and the Cauchy Integral (2002)
Vol. 1800: O. Gabber, L. Ramero, Almost Ring Theory (2003)
Vol. 1801: J. Azéma, M. Émery, M. Ledoux, M. Yor (Eds.), Séminaire de Probabilités XXXVI (2003)
Vol. 1802: V. Capasso, E. Merzbach, B. G. Ivanoff, M. Dozzi, R. Dalang, T. Mountford, Topics in Spatial Stochastic Processes. Martina Franca, Italy 2001. Editor: E. Merzbach (2003)
Vol. 1803: G. Dolzmann, Variational Methods for Crystalline Microstructure – Analysis and Computation (2003)
Vol. 1804: I. Cherednik, Ya. Markov, R. Howe, G. Lusztig, Iwahori-Hecke Algebras and their Representation Theory. Martina Franca, Italy 1999. Editors: V. Baldoni, D. Barbasch (2003)
Vol. 1805: F. Cao, Geometric Curve Evolution and Image Processing (2003)
Vol. 1806: H. Broer, I. Hoveijn. G. Lunther, G. Vegter, Bifurcations in Hamiltonian Systems. Computing Singularities by Gröbner Bases (2003)
Vol. 1807: V. D. Milman, G. Schechtman (Eds.), Geometric Aspects of Functional Analysis. Israel Seminar 2000-2002 (2003)
Vol. 1808: W. Schindler, Measures with Symmetry Properties (2003)
Vol. 1809: O. Steinbach, Stability Estimates for Hybrid Coupled Domain Decomposition Methods (2003)
Vol. 1810: J. Wengenroth, Derived Functors in Functional Analysis (2003)
Vol. 1811: J. Stevens, Deformations of Singularities (2003)
Vol. 1812: L. Ambrosio, K. Deckelnick, G. Dziuk, M. Mimura, V. A. Solonnikov, H. M. Soner, Mathematical Aspects of Evolving Interfaces. Madeira, Funchal, Portugal 2000. Editors: P. Colli, J. F. Rodrigues (2003)
Vol. 1813: L. Ambrosio, L. A. Caffarelli, Y. Brenier, G. Buttazzo, C. Villani, Optimal Transportation and its Applications. Martina Franca, Italy 2001. Editors: L. A. Caffarelli, S. Salsa (2003)
Vol. 1814: P. Bank, F. Baudoin, H. Föllmer, L.C.G. Rogers, M. Soner, N. Touzi, Paris-Princeton Lectures on Mathematical Finance 2002 (2003)
Vol. 1815: A. M. Vershik (Ed.), Asymptotic Combinatorics with Applications to Mathematical Physics. St. Petersburg, Russia 2001 (2003)
Vol. 1816: S. Albeverio, W. Schachermayer, M. Talagrand, Lectures on Probability Theory and Statistics. Ecole d'Eté de Probabilités de Saint-Flour XXX-2000. Editor: P. Bernard (2003)
Vol. 1817: E. Koelink, W. Van Assche (Eds.), Orthogonal Polynomials and Special Functions. Leuven 2002 (2003)
Vol. 1818: M. Bildhauer, Convex Variational Problems with Linear, nearly Linear and/or Anisotropic Growth Conditions (2003)
Vol. 1819: D. Masser, Yu. V. Nesterenko, H. P. Schlickewei, W. M. Schmidt, M. Waldschmidt, Diophantine Approximation. Cetraro, Italy 2000. Editors: F. Amoroso, U. Zannier (2003)
Vol. 1820: F. Hiai, H. Kosaki, Means of Hilbert Space Operators (2003)
Vol. 1821: S. Teufel, Adiabatic Perturbation Theory in Quantum Dynamics (2003)
Vol. 1822: S.-N. Chow, R. Conti, R. Johnson, J. Mallet-Paret, R. Nussbaum, Dynamical Systems. Cetraro, Italy 2000. Editors: J. W. Macki, P. Zecca (2003)
Vol. 1823: A. M. Anile, W. Allegretto, C. Ringhofer, Mathematical Problems in Semiconductor Physics. Cetraro, Italy 1998. Editor: A. M. Anile (2003)
Vol. 1824: J. A. Navarro González, J. B. Sancho de Salas, $\mathscr{C}^\infty$ – Differentiable Spaces (2003)

Vol. 1825: J. H. Bramble, A. Cohen, W. Dahmen, Multiscale Problems and Methods in Numerical Simulations, Martina Franca, Italy 2001. Editor: C. Canuto (2003)

Vol. 1826: K. Dohmen, Improved Bonferroni Inequalities via Abstract Tubes. Inequalities and Identities of Inclusion-Exclusion Type. VIII, 113 p, 2003.

Vol. 1827: K. M. Pilgrim, Combinations of Complex Dynamical Systems. IX, 118 p, 2003.

Vol. 1828: D. J. Green, Gröbner Bases and the Computation of Group Cohomology. XII, 138 p, 2003.

Vol. 1829: E. Altman, B. Gaujal, A. Hordijk, Discrete-Event Control of Stochastic Networks: Multimodularity and Regularity. XIV, 313 p, 2003.

Vol. 1830: M. I. Gil', Operator Functions and Localization of Spectra. XIV, 256 p, 2003.

Vol. 1831: A. Connes, J. Cuntz, E. Guentner, N. Higson, J. E. Kaminker, Noncommutative Geometry, Martina Franca, Italy 2002. Editors: S. Doplicher, L. Longo (2004)

Vol. 1832: J. Azéma, M. Émery, M. Ledoux, M. Yor (Eds.), Séminaire de Probabilités XXXVII (2003)

Vol. 1833: D.-Q. Jiang, M. Qian, M.-P. Qian, Mathematical Theory of Nonequilibrium Steady States. On the Frontier of Probability and Dynamical Systems. IX, 280 p, 2004.

Vol. 1834: Yo. Yomdin, G. Comte, Tame Geometry with Application in Smooth Analysis. VIII, 186 p, 2004.

Vol. 1835: O.T. Izhboldin, B. Kahn, N.A. Karpenko, A. Vishik, Geometric Methods in the Algebraic Theory of Quadratic Forms. Summer School, Lens, 2000. Editor: J.-P. Tignol (2004)

Vol. 1836: C. Năstăsescu, F. Van Oystaeyen, Methods of Graded Rings. XIII, 304 p, 2004.

Vol. 1837: S. Tavaré, O. Zeitouni, Lectures on Probability Theory and Statistics. Ecole d'Eté de Probabilités de Saint-Flour XXXI-2001. Editor: J. Picard (2004)

Vol. 1838: A.J. Ganesh, N.W. O'Connell, D.J. Wischik, Big Queues. XII, 254 p, 2004.

Vol. 1839: R. Gohm, Noncommutative Stationary Processes. VIII, 170 p, 2004.

Vol. 1840: B. Tsirelson, W. Werner, Lectures on Probability Theory and Statistics. Ecole d'Eté de Probabilités de Saint-Flour XXXII-2002. Editor: J. Picard (2004)

Vol. 1841: W. Reichel, Uniqueness Theorems for Variational Problems by the Method of Transformation Groups (2004)

Vol. 1842: T. Johnsen, A. L. Knutsen, K_3 Projective Models in Scrolls (2004)

Vol. 1843: B. Jefferies, Spectral Properties of Noncommuting Operators (2004)

Vol. 1844: K.F. Siburg, The Principle of Least Action in Geometry and Dynamics (2004)

Vol. 1845: Min Ho Lee, Mixed Automorphic Forms, Torus Bundles, and Jacobi Forms (2004)

Vol. 1846: H. Ammari, H. Kang, Reconstruction of Small Inhomogeneities from Boundary Measurements (2004)

Vol. 1847: T.R. Bielecki, T. Björk, M. Jeanblanc, M. Rutkowski, J.A. Scheinkman, W. Xiong, Paris-Princeton Lectures on Mathematical Finance 2003 (2004)

Vol. 1848: M. Abate, J. E. Fornaess, X. Huang, J. P. Rosay, A. Tumanov, Real Methods in Complex and CR Geometry, Martina Franca, Italy 2002. Editors: D. Zaitsev, G. Zampieri (2004)

Vol. 1849: Martin L. Brown, Heegner Modules and Elliptic Curves (2004)

Vol. 1850: V. D. Milman, G. Schechtman (Eds.), Geometric Aspects of Functional Analysis. Israel Seminar 2002-2003 (2004)

Vol. 1851: O. Catoni, Statistical Learning Theory and Stochastic Optimization (2004)

Vol. 1852: A.S. Kechris, B.D. Miller, Topics in Orbit Equivalence (2004)

Vol. 1853: Ch. Favre, M. Jonsson, The Valuative Tree (2004)

Vol. 1854: O. Saeki, Topology of Singular Fibers of Differential Maps (2004)

Vol. 1855: G. Da Prato, P.C. Kunstmann, I. Lasiecka, A. Lunardi, R. Schnaubelt, L. Weis, Functional Analytic Methods for Evolution Equations. Editors: M. Iannelli, R. Nagel, S. Piazzera (2004)

Vol. 1856: K. Back, T.R. Bielecki, C. Hipp, S. Peng, W. Schachermayer, Stochastic Methods in Finance, Bressanone/Brixen, Italy, 2003. Editors: M. Fritelli, W. Runggaldier (2004)

Vol. 1857: M. Émery, M. Ledoux, M. Yor (Eds.), Séminaire de Probabilités XXXVIII (2005)

Vol. 1858: A.S. Cherny, H.-J. Engelbert, Singular Stochastic Differential Equations (2005)

Vol. 1859: E. Letellier, Fourier Transforms of Invariant Functions on Finite Reductive Lie Algebras (2005)

Vol. 1860: A. Borisyuk, G.B. Ermentrout, A. Friedman, D. Terman, Tutorials in Mathematical Biosciences I. Mathematical Neurosciences (2005)

Vol. 1861: G. Benettin, J. Henrard, S. Kuksin, Hamiltonian Dynamics – Theory and Applications, Cetraro, Italy, 1999. Editor: A. Giorgilli (2005)

Vol. 1862: B. Helffer, F. Nier, Hypoelliptic Estimates and Spectral Theory for Fokker-Planck Operators and Witten Laplacians (2005)

Vol. 1863: H. Führ, Abstract Harmonic Analysis of Continuous Wavelet Transforms (2005)

Vol. 1864: K. Efstathiou, Metamorphoses of Hamiltonian Systems with Symmetries (2005)

Vol. 1865: D. Applebaum, B.V. R. Bhat, J. Kustermans, J. M. Lindsay, Quantum Independent Increment Processes I. From Classical Probability to Quantum Stochastic Calculus. Editors: M. Schürmann, U. Franz (2005)

Vol. 1866: O.E. Barndorff-Nielsen, U. Franz, R. Gohm, B. Kümmerer, S. Thorbjønsen, Quantum Independent Increment Processes II. Structure of Quantum Lévy Processes, Classical Probability, and Physics. Editors: M. Schürmann, U. Franz, (2005)

Vol. 1867: J. Sneyd (Ed.), Tutorials in Mathematical Biosciences II. Mathematical Modeling of Calcium Dynamics and Signal Transduction. (2005)

Vol. 1868: J. Jorgenson, S. Lang, $Pos_n(R)$ and Eisenstein Series. (2005)

Vol. 1869: A. Dembo, T. Funaki, Lectures on Probability Theory and Statistics. Ecole d'Eté de Probabilités de Saint-Flour XXXIII-2003. Editor: J. Picard (2005)

Vol. 1870: V.I. Gurariy, W. Lusky, Geometry of Müntz Spaces and Related Questions. (2005)

Vol. 1871: P. Constantin, G. Gallavotti, A.V. Kazhikhov, Y. Meyer, S. Ukai, Mathematical Foundation of Turbulent Viscous Flows, Martina Franca, Italy, 2003. Editors: M. Cannone, T. Miyakawa (2006)

Vol. 1872: A. Friedman (Ed.), Tutorials in Mathematical Biosciences III. Cell Cycle, Proliferation, and Cancer (2006)

Vol. 1873: R. Mansuy, M. Yor, Random Times and Enlargements of Filtrations in a Brownian Setting (2006)

Vol. 1874: M. Yor, M. Émery (Eds.), In Memoriam Paul-André Meyer - Séminaire de Probabilités XXXIX (2006)
Vol. 1875: J. Pitman, Combinatorial Stochastic Processes. Ecole d'Eté de Probabilités de Saint-Flour XXXII-2002. Editor: J. Picard (2006)
Vol. 1876: H. Herrlich, Axiom of Choice (2006)
Vol. 1877: J. Steuding, Value Distributions of L-Functions (2007)
Vol. 1878: R. Cerf, The Wulff Crystal in Ising and Percolation Models, Ecole d'Eté de Probabilités de Saint-Flour XXXIV-2004. Editor: Jean Picard (2006)
Vol. 1879: G. Slade, The Lace Expansion and its Applications, Ecole d'Eté de Probabilités de Saint-Flour XXXIV-2004. Editor: Jean Picard (2006)
Vol. 1880: S. Attal, A. Joye, C.-A. Pillet, Open Quantum Systems I, The Hamiltonian Approach (2006)
Vol. 1881: S. Attal, A. Joye, C.-A. Pillet, Open Quantum Systems II, The Markovian Approach (2006)
Vol. 1882: S. Attal, A. Joye, C.-A. Pillet, Open Quantum Systems III, Recent Developments (2006)
Vol. 1883: W. Van Assche, F. Marcellàn (Eds.), Orthogonal Polynomials and Special Functions, Computation and Application (2006)
Vol. 1884: N. Hayashi, E.I. Kaikina, P.I. Naumkin, I.A. Shishmarev, Asymptotics for Dissipative Nonlinear Equations (2006)
Vol. 1885: A. Telcs, The Art of Random Walks (2006)
Vol. 1886: S. Takamura, Splitting Deformations of Degenerations of Complex Curves (2006)
Vol. 1887: K. Habermann, L. Habermann, Introduction to Symplectic Dirac Operators (2006)
Vol. 1888: J. van der Hoeven, Transseries and Real Differential Algebra (2006)
Vol. 1889: G. Osipenko, Dynamical Systems, Graphs, and Algorithms (2006)
Vol. 1890: M. Bunge, J. Funk, Singular Coverings of Toposes (2006)
Vol. 1891: J.B. Friedlander, D.R. Heath-Brown, H. Iwaniec, J. Kaczorowski, Analytic Number Theory, Cetraro, Italy, 2002. Editors: A. Perelli, C. Viola (2006)
Vol. 1892: A. Baddeley, I. Bárány, R. Schneider, W. Weil, Stochastic Geometry, Martina Franca, Italy, 2004. Editor: W. Weil (2007)
Vol. 1893: H. Hanßmann, Local and Semi-Local Bifurcations in Hamiltonian Dynamical Systems, Results and Examples (2007)
Vol. 1894: C.W. Groetsch, Stable Approximate Evaluation of Unbounded Operators (2007)
Vol. 1895: L. Molnár, Selected Preserver Problems on Algebraic Structures of Linear Operators and on Function Spaces (2007)
Vol. 1896: P. Massart, Concentration Inequalities and Model Selection, Ecole d'Été de Probabilités de Saint-Flour XXXIII-2003. Editor: J. Picard (2007)
Vol. 1897: R. Doney, Fluctuation Theory for Lévy Processes, Ecole d'Été de Probabilités de Saint-Flour XXXV-2005. Editor: J. Picard (2007)
Vol. 1898: H.R. Beyer, Beyond Partial Differential Equations, On linear and Quasi-Linear Abstract Hyperbolic Evolution Equations (2007)
Vol. 1899: Séminaire de Probabilités XL. Editors: C. Donati-Martin, M. Émery, A. Rouault, C. Stricker (2007)
Vol. 1900: E. Bolthausen, A. Bovier (Eds.), Spin Glasses (2007)
Vol. 1901: O. Wittenberg, Intersections de deux quadriques et pinceaux de courbes de genre 1, Intersections of Two Quadrics and Pencils of Curves of Genus 1 (2007)
Vol. 1902: A. Isaev, Lectures on the Automorphism Groups of Kobayashi-Hyperbolic Manifolds (2007)
Vol. 1903: G. Kresin, V. Maz'ya, Sharp Real-Part Theorems (2007)
Vol. 1904: P. Giesl, Construction of Global Lyapunov Functions Using Radial Basis Functions (2007)
Vol. 1905: C. Prévôt, M. Röckner, A Concise Course on Stochastic Partial Differential Equations (2007)
Vol. 1906: T. Schuster, The Method of Approximate Inverse: Theory and Applications (2007)
Vol. 1907: M. Rasmussen, Attractivity and Bifurcation for Nonautonomous Dynamical Systems (2007)
Vol. 1908: T.J. Lyons, M. Caruana, T. Lévy, Differential Equations Driven by Rough Paths, Ecole d'Été de Probabilités de Saint-Flour XXXIV-2004 (2007)
Vol. 1909: H. Akiyoshi, M. Sakuma, M. Wada, Y. Yamashita, Punctured Torus Groups and 2-Bridge Knot Groups (I) (2007)
Vol. 1910: V.D. Milman, G. Schechtman (Eds.), Geometric Aspects of Functional Analysis. Israel Seminar 2004-2005 (2007)
Vol. 1911: A. Bressan, D. Serre, M. Williams, K. Zumbrun, Hyperbolic Systems of Balance Laws. Cetraro, Italy 2003. Editor: P. Marcati (2007)
Vol. 1912: V. Berinde, Iterative Approximation of Fixed Points (2007)
Vol. 1913: J.E. Marsden, G. Misiołek, J.-P. Ortega, M. Perlmutter, T.S. Ratiu, Hamiltonian Reduction by Stages (2007)
Vol. 1914: G. Kutyniok, Affine Density in Wavelet Analysis (2007)
Vol. 1915: T. Bıyıkoğlu, J. Leydold, P.F. Stadler, Laplacian Eigenvectors of Graphs. Perron-Frobenius and Faber-Krahn Type Theorems (2007)
Vol. 1916: C. Villani, F. Rezakhanlou, Entropy Methods for the Boltzmann Equation. Editors: F. Golse, S. Olla (2008)
Vol. 1917: I. Veselić, Existence and Regularity Properties of the Integrated Density of States of Random Schrödinger (2008)
Vol. 1918: B. Roberts, R. Schmidt, Local Newforms for GSp(4) (2007)
Vol. 1919: R.A. Carmona, I. Ekeland, A. Kohatsu-Higa, J.-M. Lasry, P.-L. Lions, H. Pham, E. Taflin, Paris-Princeton Lectures on Mathematical Finance 2004. Editors: R.A. Carmona, E. Çinlar, I. Ekeland, E. Jouini, J.A. Scheinkman, N. Touzi (2007)
Vol. 1920: S.N. Evans, Probability and Real Trees. Ecole d'Été de Probabilités de Saint-Flour XXXV-2005 (2008)
Vol. 1921: J.P. Tian, Evolution Algebras and their Applications (2008)
Vol. 1922: A. Friedman (Ed.), Tutorials in Mathematical BioSciences IV. Evolution and Ecology (2008)
Vol. 1923: J.P.N. Bishwal, Parameter Estimation in Stochastic Differential Equations (2008)
Vol. 1924: M. Wilson, Littlewood-Paley Theory and Exponential-Square Integrability (2008)
Vol. 1925: M. du Sautoy, L. Woodward, Zeta Functions of Groups and Rings (2008)
Vol. 1926: L. Barreira, V. Claudia, Stability of Nonautonomous Differential Equations (2008)
Vol. 1927: L. Ambrosio, L. Caffarelli, M.G. Crandall, L.C. Evans, N. Fusco, Calculus of Variations and Non-Linear Partial Differential Equations. Cetraro, Italy 2005. Editors: B. Dacorogna, P. Marcellini (2008)

Vol. 1928: J. Jonsson, Simplicial Complexes of Graphs (2008)
Vol. 1929: Y. Mishura, Stochastic Calculus for Fractional Brownian Motion and Related Processes (2008)
Vol. 1930: J.M. Urbano, The Method of Intrinsic Scaling. A Systematic Approach to Regularity for Degenerate and Singular PDEs (2008)
Vol. 1931: M. Cowling, E. Frenkel, M. Kashiwara, A. Valette, D.A. Vogan, Jr., N.R. Wallach, Representation Theory and Complex Analysis. Venice, Italy 2004. Editors: E.C. Tarabusi, A. D'Agnolo, M. Picardello (2008)
Vol. 1932: A.A. Agrachev, A.S. Morse, E.D. Sontag, H.J. Sussmann, V.I. Utkin, Nonlinear and Optimal Control Theory. Cetraro, Italy 2004. Editors: P. Nistri, G. Stefani (2008)
Vol. 1933: M. Petkovic, Point Estimation of Root Finding Methods (2008)
Vol. 1934: C. Donati-Martin, M. Émery, A. Rouault, C. Stricker (Eds.), Séminaire de Probabilités XLI (2008)
Vol. 1935: A. Unterberger, Alternative Pseudodifferential Analysis (2008)
Vol. 1936: P. Magal, S. Ruan (Eds.), Structured Population Models in Biology and Epidemiology (2008)
Vol. 1937: G. Capriz, P. Giovine, P.M. Mariano (Eds.), Mathematical Models of Granular Matter (2008)
Vol. 1938: D. Auroux, F. Catanese, M. Manetti, P. Seidel, B. Siebert, I. Smith, G. Tian, Symplectic 4-Manifolds and Algebraic Surfaces. Cetraro, Italy 2003. Editors: F. Catanese, G. Tian (2008)
Vol. 1939: D. Boffi, F. Brezzi, L. Demkowicz, R.G. Durán, R.S. Falk, M. Fortin, Mixed Finite Elements, Compatibility Conditions, and Applications. Cetraro, Italy 2006. Editors: D. Boffi, L. Gastaldi (2008)
Vol. 1940: J. Banasiak, V. Capasso, M.A.J. Chaplain, M. Lachowicz, J. Miękisz, Multiscale Problems in the Life Sciences. From Microscopic to Macroscopic. Będlewo, Poland 2006. Editors: V. Capasso, M. Lachowicz (2008)
Vol. 1941: S.M.J. Haran, Arithmetical Investigations. Representation Theory, Orthogonal Polynomials, and Quantum Interpolations (2008)
Vol. 1942: S. Albeverio, F. Flandoli, Y.G. Sinai, SPDE in Hydrodynamic. Recent Progress and Prospects. Cetraro, Italy 2005. Editors: G. Da Prato, M. Röckner (2008)
Vol. 1943: L.L. Bonilla (Ed.), Inverse Problems and Imaging. Martina Franca, Italy 2002 (2008)
Vol. 1944: A. Di Bartolo, G. Falcone, P. Plaumann, K. Strambach, Algebraic Groups and Lie Groups with Few Factors (2008)
Vol. 1945: F. Brauer, P. van den Driessche, J. Wu (Eds.), Mathematical Epidemiology (2008)
Vol. 1946: G. Allaire, A. Arnold, P. Degond, T.Y. Hou, Quantum Transport. Modelling, Analysis and Asymptotics. Cetraro, Italy 2006. Editors: N.B. Abdallah, G. Frosali (2008)
Vol. 1947: D. Abramovich, M. Mariño, M. Thaddeus, R. Vakil, Enumerative Invariants in Algebraic Geometry and String Theory. Cetraro, Italy 2005. Editors: K. Behrend, M. Manetti (2008)
Vol. 1948: F. Cao, J-L. Lisani, J-M. Morel, P. Musé, F. Sur, A Theory of Shape Identification (2008)
Vol. 1949: H.G. Feichtinger, B. Helffer, M.P. Lamoureux, N. Lerner, J. Toft, Pseudo-Differential Operators. Quantization and Signals. Cetraro, Italy 2006. Editors: L. Rodino, M.W. Wong (2008)
Vol. 1950: M. Bramson, Stability of Queueing Networks, Ecole d'Eté de Probabilités de Saint-Flour XXXVI-2006 (2008)
Vol. 1951: A. Moltó, J. Orihuela, S. Troyanski, M. Valdivia, A Non Linear Transfer Technique for Renorming (2009)
Vol. 1952: R. Mikhailov, I.B.S. Passi, Lower Central and Dimension Series of Groups (2009)
Vol. 1953: K. Arwini, C.T.J. Dodson, Information Geometry (2008)
Vol. 1954: P. Biane, L. Bouten, F. Cipriani, N. Konno, N. Privault, Q. Xu, Quantum Potential Theory. Editors: U. Franz, M. Schuermann (2008)
Vol. 1955: M. Bernot, V. Caselles, J.-M. Morel, Optimal Transportation Networks (2009)
Vol. 1956: C.H. Chu, Matrix Convolution Operators on Groups (2008)
Vol. 1957: A. Guionnet, On Random Matrices: Macroscopic Asymptotics, Ecole d'Eté de Probabilités de Saint-Flour XXXVI-2006 (2009)
Vol. 1958: M.C. Olsson, Compactifying Moduli Spaces for Abelian Varieties (2008)
Vol. 1959: Y. Nakkajima, A. Shiho, Weight Filtrations on Log Crystalline Cohomologies of Families of Open Smooth Varieties (2008)
Vol. 1960: J. Lipman, M. Hashimoto, Foundations of Grothendieck Duality for Diagrams of Schemes (2009)
Vol. 1961: G. Buttazzo, A. Pratelli, S. Solimini, E. Stepanov, Optimal Urban Networks via Mass Transportation (2009)
Vol. 1962: R. Dalang, D. Khoshnevisan, C. Mueller, D. Nualart, Y. Xiao, A Minicourse on Stochastic Partial Differential Equations (2009)
Vol. 1963: W. Siegert, Local Lyapunov Exponents (2009)
Vol. 1964: W. Roth, Operator-valued Measures and Integrals for Cone-valued Functions and Integrals for Cone-valued Functions (2009)
Vol. 1965: C. Chidume, Geometric Properties of Banach Spaces and Nonlinear Iterations (2009)
Vol. 1966: D. Deng, Y. Han, Harmonic Analysis on Spaces of Homogeneous Type (2009)

Recent Reprints and New Editions

Vol. 1702: J. Ma, J. Yong, Forward-Backward Stochastic Differential Equations and their Applications. 1999 – Corr. 3rd printing (2007)
Vol. 830: J.A. Green, Polynomial Representations of GL_n, with an Appendix on Schensted Correspondence and Littelmann Paths by K. Erdmann, J.A. Green and M. Schoker 1980 – 2nd corr. and augmented edition (2007)
Vol. 1693: S. Simons, From Hahn-Banach to Monotonicity (Minimax and Monotonicity 1998) – 2nd exp. edition (2008)
Vol. 470: R.E. Bowen, Equilibrium States and the Ergodic Theory of Anosov Diffeomorphisms. With a preface by D. Ruelle. Edited by J.-R. Chazottes. 1975 – 2nd rev. edition (2008)
Vol. 523: S.A. Albeverio, R.J. Høegh-Krohn, S. Mazzucchi, Mathematical Theory of Feynman Path Integral. 1976 – 2nd corr. and enlarged edition (2008)
Vol. 1764: A. Cannas da Silva, Lectures on Symplectic Geometry 2001 – Corr. 2nd printing (2008)

LECTURE NOTES IN MATHEMATICS 

Edited by J.-M. Morel, F. Takens, B. Teissier, P.K. Maini

Editorial Policy (for the publication of monographs)

1. Lecture Notes aim to report new developments in all areas of mathematics and their applications - quickly, informally and at a high level. Mathematical texts analysing new developments in modelling and numerical simulation are welcome.

   Monograph manuscripts should be reasonably self-contained and rounded off. Thus they may, and often will, present not only results of the author but also related work by other people. They may be based on specialised lecture courses. Furthermore, the manuscripts should provide sufficient motivation, examples and applications. This clearly distinguishes Lecture Notes from journal articles or technical reports which normally are very concise. Articles intended for a journal but too long to be accepted by most journals, usually do not have this "lecture notes" character. For similar reasons it is unusual for doctoral theses to be accepted for the Lecture Notes series, though habilitation theses may be appropriate.

2. Manuscripts should be submitted either to Springer's mathematics editorial in Heidelberg, or to one of the series editors. In general, manuscripts will be sent out to 2 external referees for evaluation. If a decision cannot yet be reached on the basis of the first 2 reports, further referees may be contacted: The author will be informed of this. A final decision to publish can be made only on the basis of the complete manuscript, however a refereeing process leading to a preliminary decision can be based on a pre-final or incomplete manuscript. The strict minimum amount of material that will be considered should include a detailed outline describing the planned contents of each chapter, a bibliography and several sample chapters.

   Authors should be aware that incomplete or insufficiently close to final manuscripts almost always result in longer refereeing times and nevertheless unclear referees' recommendations, making further refereeing of a final draft necessary.

   Authors should also be aware that parallel submission of their manuscript to another publisher while under consideration for LNM will in general lead to immediate rejection.

3. Manuscripts should in general be submitted in English. Final manuscripts should contain at least 100 pages of mathematical text and should always include

   – a table of contents;
   – an informative introduction, with adequate motivation and perhaps some historical remarks: it should be accessible to a reader not intimately familiar with the topic treated;
   – a subject index: as a rule this is genuinely helpful for the reader.

   For evaluation purposes, manuscripts may be submitted in print or electronic form, in the latter case preferably as pdf- or zipped ps-files. Lecture Notes volumes are, as a rule, printed digitally from the authors' files. To ensure best results, authors are asked to use the LaTeX2e style files available from Springer's web-server at:

   ftp://ftp.springer.de/pub/tex/latex/svmonot1/ (for monographs).

Additional technical instructions, if necessary, are available on request from: lnm@springer.com.

4. Careful preparation of the manuscripts will help keep production time short besides ensuring satisfactory appearance of the finished book in print and online. After acceptance of the manuscript authors will be asked to prepare the final LaTeX source files (and also the corresponding dvi-, pdf- or zipped ps-file) together with the final printout made from these files. The LaTeX source files are essential for producing the full-text online version of the book (see www.springerlink.com/content/110312 for the existing online volumes of LNM).

   The actual production of a Lecture Notes volume takes approximately 12 weeks.

5. Authors receive a total of 50 free copies of their volume, but no royalties. They are entitled to a discount of 33.3% on the price of Springer books purchased for their personal use, if ordering directly from Springer.

6. Commitment to publish is made by letter of intent rather than by signing a formal contract. Springer-Verlag secures the copyright for each volume. Authors are free to reuse material contained in their LNM volumes in later publications: a brief written (or e-mail) request for formal permission is sufficient.

Addresses:

Professor J.-M. Morel, CMLA,
École Normale Supérieure de Cachan,
61 Avenue du Président Wilson, 94235 Cachan Cedex, France
E-mail: Jean-Michel.Morel@cmla.ens-cachan.fr

Professor F. Takens, Mathematisch Instituut,
Rijksuniversiteit Groningen, Postbus 800,
9700 AV Groningen, The Netherlands
E-mail: F.Takens@math.rug.nl

Professor B. Teissier, Institut Mathématique de Jussieu,
UMR 7586 du CNRS, Équipe "Géométrie et Dynamique",
175 rue du Chevaleret
75013 Paris, France
E-mail: teissier@math.jussieu.fr

For the "Mathematical Biosciences Subseries" of LNM:

Professor P.K. Maini, Center for Mathematical Biology,
Mathematical Institute, 24-29 St Giles,
Oxford OX1 3LP, UK
E-mail: maini@maths.ox.ac.uk

Springer, Mathematics Editorial I, Tiergartenstr. 17
69121 Heidelberg, Germany,
Tel.: +49 (6221) 487-8259
Fax: +49 (6221) 4876-8259
E-mail: lnm@springer.com